LONGMAN PHYSICS General Editor: John L. Lewis

RADIOACTIVITY

John L. Lewis
*Senior Science Master, Malvern College
formerly Associate Organiser,
Nuffield O-level Physics Project*

and

E. J. Wenham
*Principal Lecturer in Physics, Worcester College of Education
formerly Associate Organiser,
Nuffield O-level Physics Project*

Illustrated by T. H. McArthur

LONGMAN

LONGMAN GROUP LIMITED
London

Associated companies, branches and representatives throughout the world

© Longman Group Ltd 1970

First published 1970
Fifth impression 1980
ISBN 0 582 32206 5

Printed in Hong Kong by Wing Tai Cheung Printing Co Ltd

NOTE TO THE TEACHER

This book is one in the series of physics background books intended primarily for use with the Nuffield O-level Physics Project. The team of writers who have contributed to the series were all associated with the Project. It was always intended that the Nuffield teachers' material should be accompanied by background books for pupils to read, and a number of such books is being produced under the Foundation's auspices. This series of books is intended as a supplement to the Nuffield pupils' material: not books giving the answers to all the investigations pupils will be doing in the laboratory, certainly not textbooks in the conventional sense, but books that are easy to read and copiously illustrated, and which show how the principles studied in school are applied in the outside world.

The books are such that they can be used with a conventional as well as a modern physics programme. Whatever course pupils are following, they often need straightforward books to help clarify their knowledge, sometimes to help them catch up on any topic they have missed in the school course. It is hoped that this series will meet that need.

This background series will provide suitable material for reading in homework. Each volume is divided into sections, and the teacher may feel that one section at a time is suitable for each homework session.

This particular book is written as a background book for the work on radioactivity in Year V of the Nuffield course, but the material is also suitable for those studying radioactivity in conventional O-level or A-level courses.

CONTENTS

Above: nuclear explosion

Below: Hunterston Nuclear Power Station

In a cloud chamber you can see something of the power of the atom: a radioactive substance throwing out a part of itself as some energy within the atom is released. This century has seen great advances in our knowledge of the atom and its nucleus, culminating in its use both for destructive purposes and as a great source of energy for the benefit of mankind.

In this book we shall discuss the experiments which you have seen at school. Then we will look at the early history of radioactivity and some of the remarkable people who contributed to that history. Brief consideration of nuclear reactions will lead us to artificial radioactivity and the many uses to which this is put at the present time.

The study of radioactivity is an exciting story starting with Becquerel's discovery. It is one which has captured the imagination of great scientists pursuing knowledge for its own sake. Madame Curie, doubtless as a consequence of her inherent modesty about herself, often said that in science we should be interested in things, not in persons. To-day that can no longer be the claim of the scientist, for he has heavy responsibilities toward people, and can use his knowledge both for good and for ill. We should all have some understanding of what radioactivity and nuclear reactions are about.

Above: vapour trails in the sky

Below: tracks in a cloud chamber

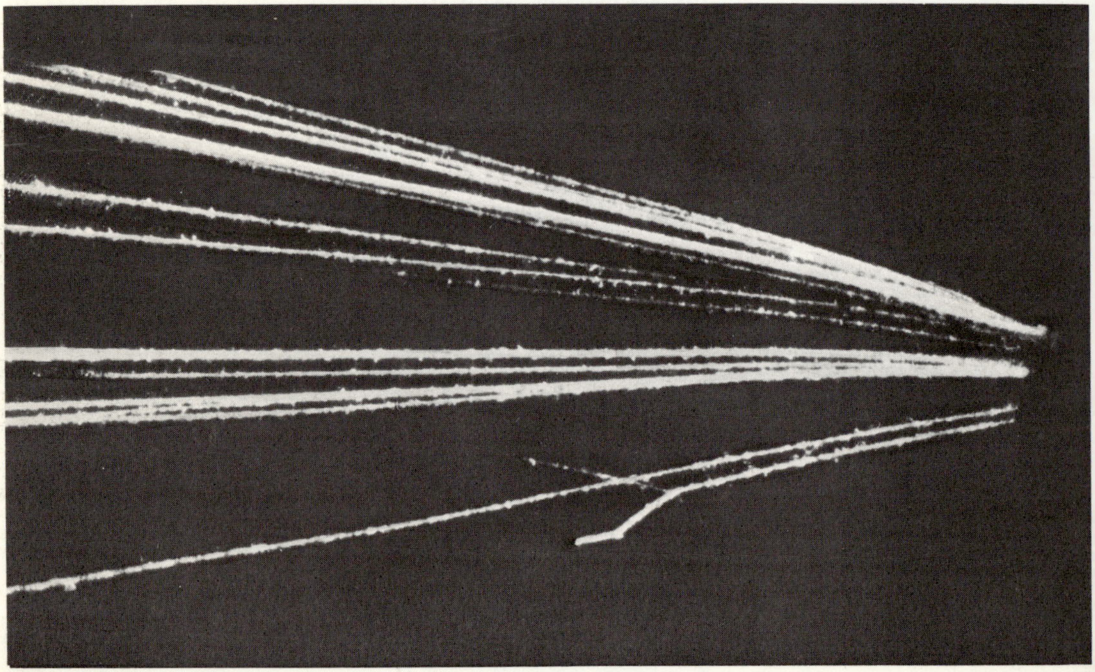

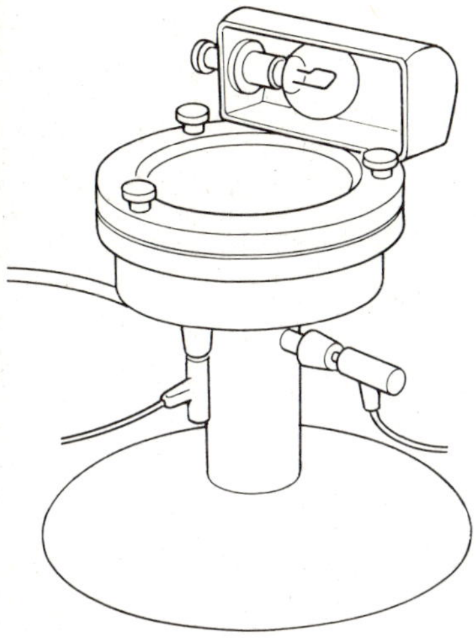

EVIDENCE FROM CLOUD CHAMBERS

In your school laboratory you have seen tracks of radio-active particles in a cloud chamber. You did not, of course, see the particles themselves any more than you can necessarily see the aircraft causing vapour trails high in the sky; in the laboratory you saw where the particles went, and in the sky where the aircraft went.

The cloud chamber is a powerful aid in helping us in our study of radioactivity and you will find many photographs of cloud-chamber tracks in this book. There are two types of cloud chamber: the expansion type, first developed by C. T. R. Wilson, and the continuous cloud chamber, first suggested by Langsdorf in 1936. You have probably seen both in your school. A typical expansion cloud chamber for use in schools is shown on the left.

In both cloud chambers the tracks are caused by the condensation of liquid droplets. This is also the cause of vapour trails in the sky. How the cloud chamber works will be discussed later on page 25, but here are some typical cloud-chamber photographs.

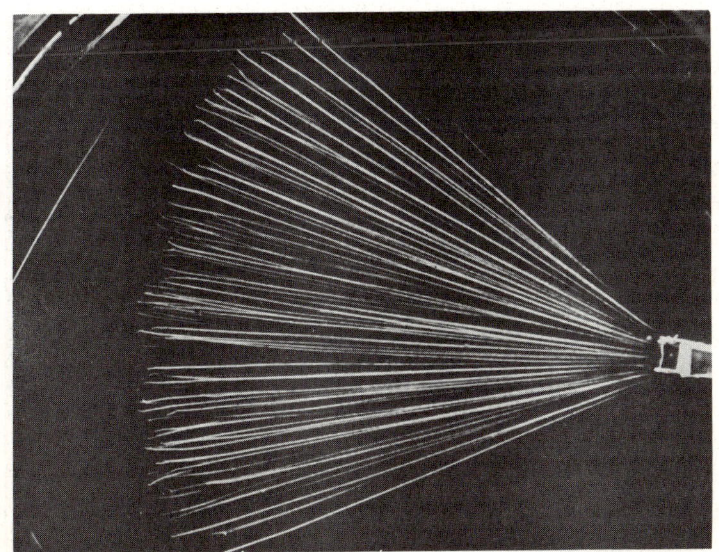

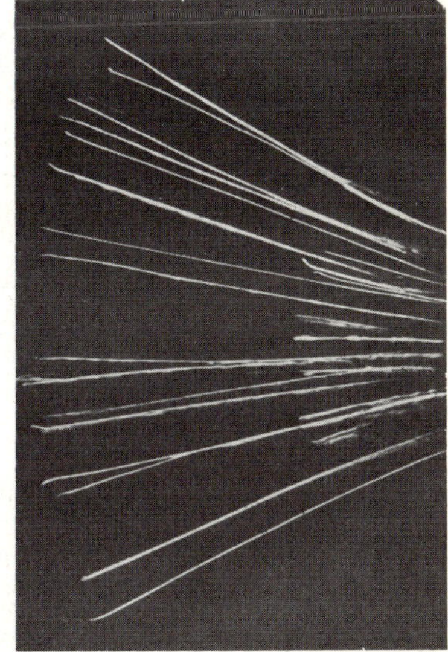

* You will find notes on these numbered questions on page 61, but think about them before looking.

What do you notice about the range of the alpha particles in the first photograph?[1] *What is the difference between the first and second photographs?[2]*

The next two photographs were taken when some radioactive gas, thoron, was put into the cloud chamber. Because the activity is not concentrated in one place, the tracks appear all over the chamber.

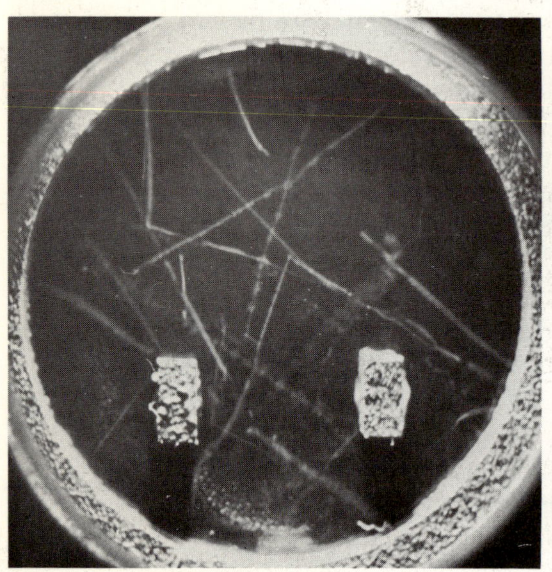 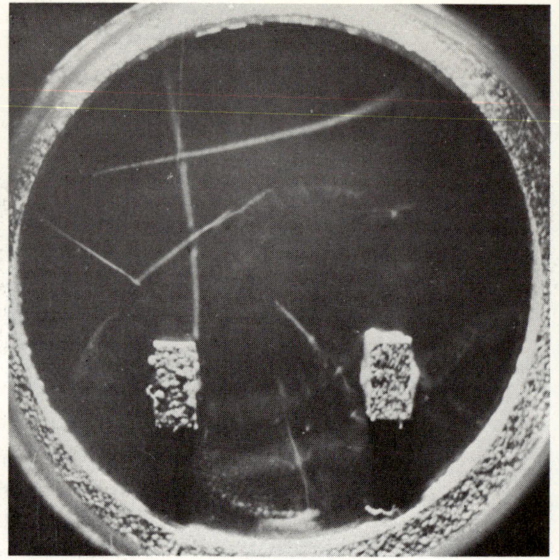

The two photographs were taken two minutes apart. *What do you notice about the difference between the two?*[3]

The photographs show typical alpha-particle tracks. The tracks are made by the particles, which create ions as they pass through the chamber. On these ions some vapour condenses. In other words, the tracks are made up of individual droplets. This process will be discussed more fully in a later chapter.

All the photographs shown so far have been of one type. It is interesting to compare the four photographs on the opposite page. *What differences do you notice?*[4]

The second and third photographs are somewhat similar in appearance: one shows a high-speed particle and the other a similar particle, moving at a much slower speed and so more easily knocked from a straight-line course.

The fourth photograph was made when gamma rays entered the chamber on the right. Such rays produce very few ions, but the tracks of secondary electrons ejected from the air can be seen.

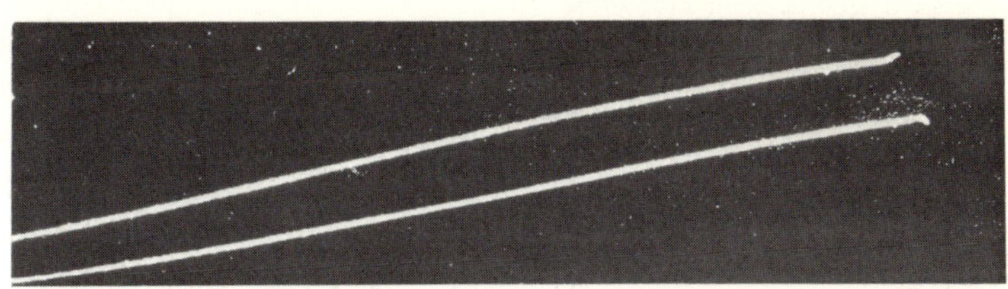

Alpha track

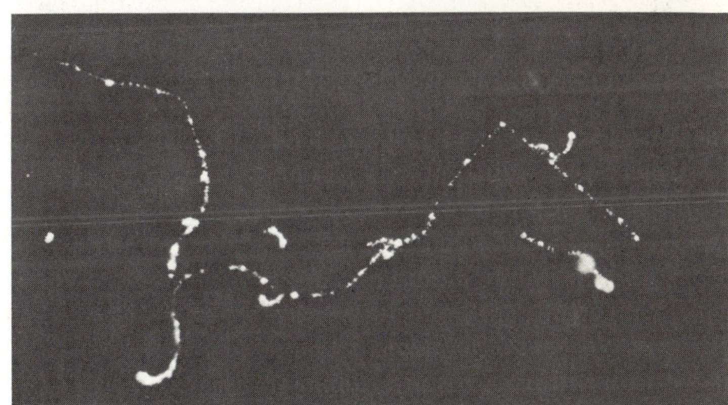

Straight beta track (high energy)

Irregular beta tracks (low energy)

*Gamma track showing secondary
electrons (gamma rays come from
the direction of the arrow)*

Photographs such as these suggest the existence of three different kinds of radiation. As you have heard in school and as will be discussed on page 31, these were called alpha, beta and gamma radiation. The alpha particle produces large numbers of ions on its journey, and so a large number of droplets can form, condensing on the ions, and you see a dense track. The beta particles produce far fewer ions and so the tracks are much less dense.

Below and right
Beta tracks showing pairs of droplets

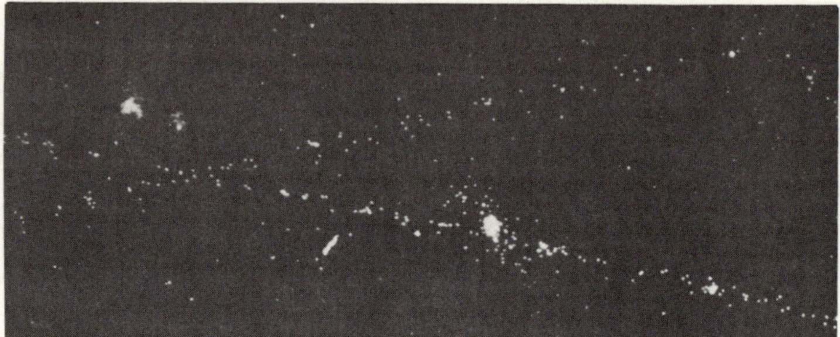

The pictures shown here are interesting. They are magnified photographs of beta-particle tracks. You can see in the photographs that the droplets form in pairs. When ionisation occurs, an electron is released from an atom or molecule, leaving a positive ion. This electron usually attaches itself quickly to a neutral molecule. This process means that both a positive and a negative molecular ion are provided on which water drops condense. This explains why the droplets form in pairs.

Another interesting property of certain particles is shown in the photograph below. The particles are moving

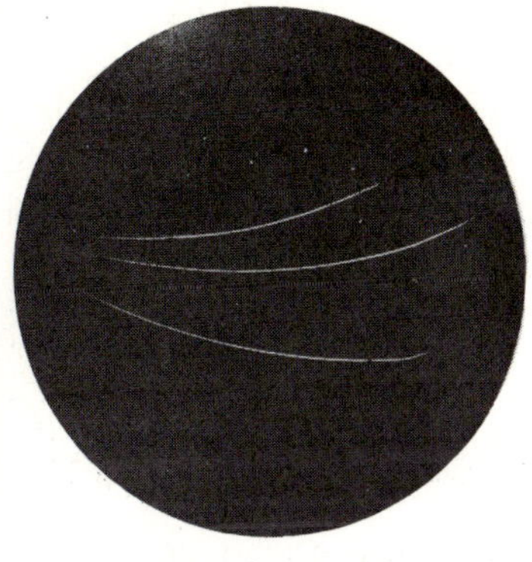

Deflection of alpha particles in strong magnetic field

in a magnetic field and are noticeably affected by it. The beta particle can easily be deflected into a curved path.

In fact alpha particles *can* be deflected in a magnetic field, but it requires a very much stronger field to do so. The photograph on the left shows alpha particles being deflected in this way.

It is found that alpha and beta particles are deflected in opposite directions. *Why do you think this is?*[5] How much a particle is deflected will obviously depend on the mass of the particle and on the strength of the field. It also depends on how fast the particle is moving. *Would a fast-moving particle be deflected more or less than a slow-moving one?*[6]

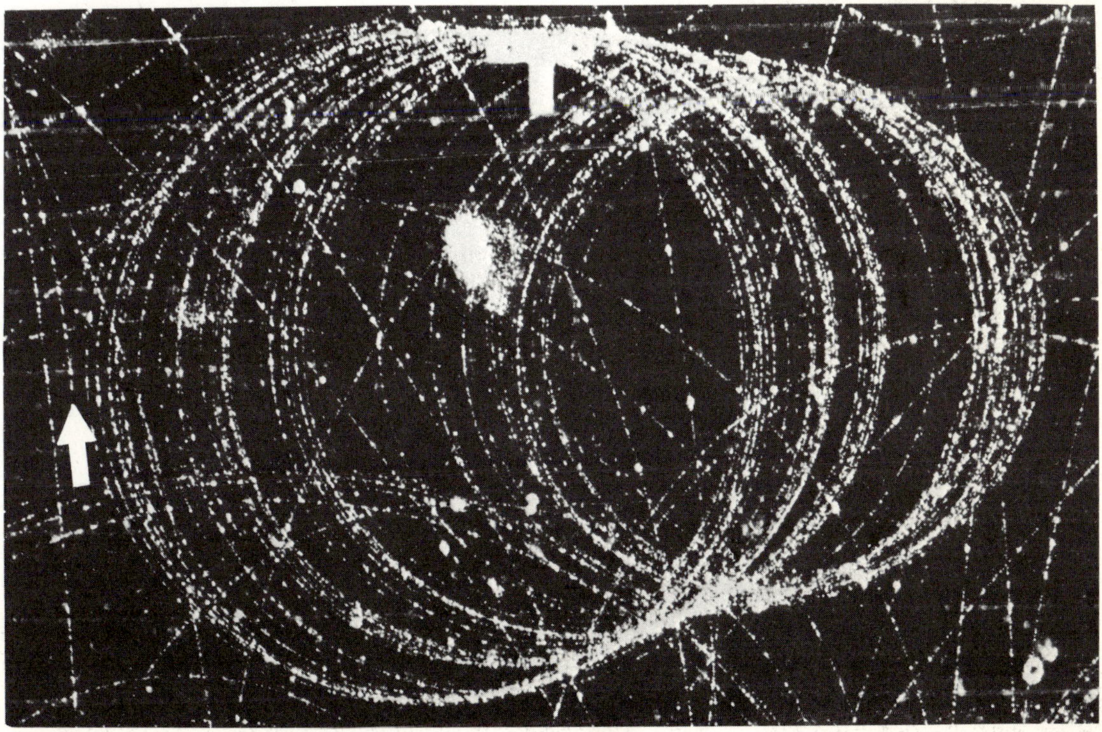

The picture below is a remarkable one of a beta particle (or electron) in a magnetic field. It was taken at the Radiation Laboratory, Berkeley, California. The electron was produced at the point marked with a white arrow and it spirals thirty-six times in the magnetic field. The field was not quite uniform, which explains the drift of the circles to the right, fortunately making it easier for us to see. Toward the end of its path, the diameter of the circles becomes less, giving the spiralling effect. *Why do you think this happens?*[7] You can also see other tracks crossing the photograph. *Do these tracks look like alpha particles or beta particles?*[8] *Have they higher or lower energy than the electron in the spiral?*[9]

Cloud-chamber photographs lead us to believe that there are three different kinds of radiation from radioactive substances. Further evidence for this comes from other experiments which you have seen in your school laboratory.

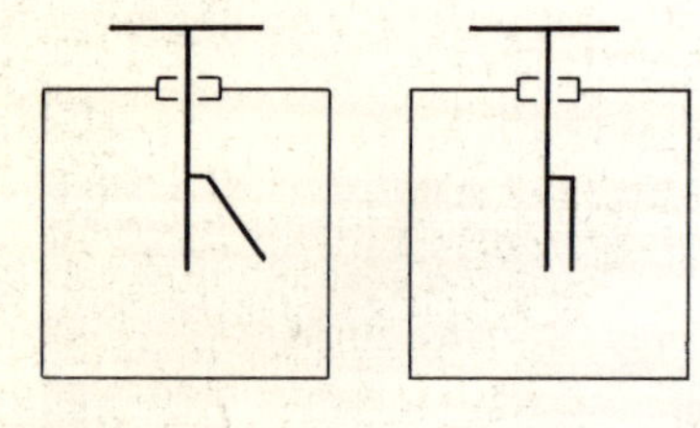

ELECTROSCOPE AND SPARK COUNTER

One method of detection is to use a gold-leaf electroscope. When an electroscope is charged, the leaf rises; but if a lighted match is held near it the electroscope discharges and the leaf falls; similarly, if a radium source is held near it, the electroscope discharges and the leaf falls steadily. You know already that such a source produces ions and these ions cause the charge to leak away.

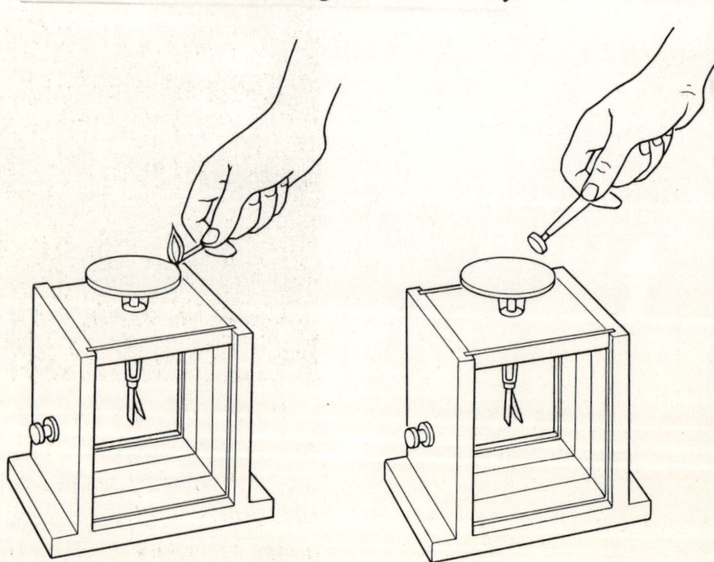

Another method is to use a spark counter. In this a high voltage, usually between 4000 and 5000 volts, is applied to a fine wire at a fixed distance below a wire gauze, which is earthed. The voltage is just too small for a spark to pass between the wire and the gauze. (It is usually set up by increasing the voltage until sparking starts and then reducing the voltage very slightly until it stops.) A lighted match produces ions; if a match is held near the counter and you blow gently, the ions start a spark. In the same way, a radium source held over the spark counter produces ions between the gauze and the wire and causes a spark. The stronger the radioactive source the more sparks there are.

A spark counter in use

When you saw the experiment, what happened when the source was raised higher and higher above the spark counter?[10] *What happened when a piece of paper was put between the source and the spark counter?*[11] *What did you notice about the sparks? Were they regular?*[12]

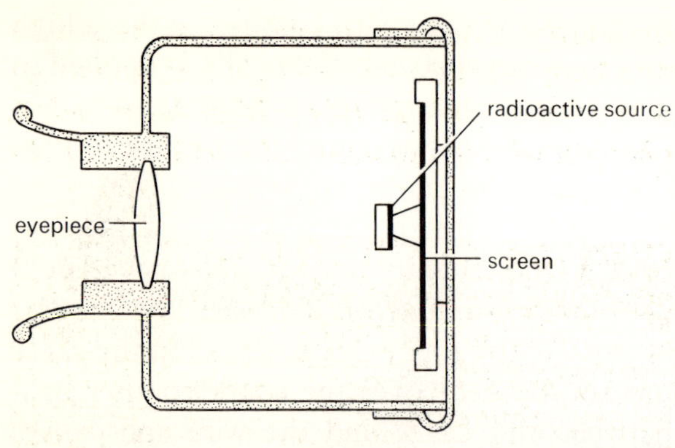

A spinthariscope

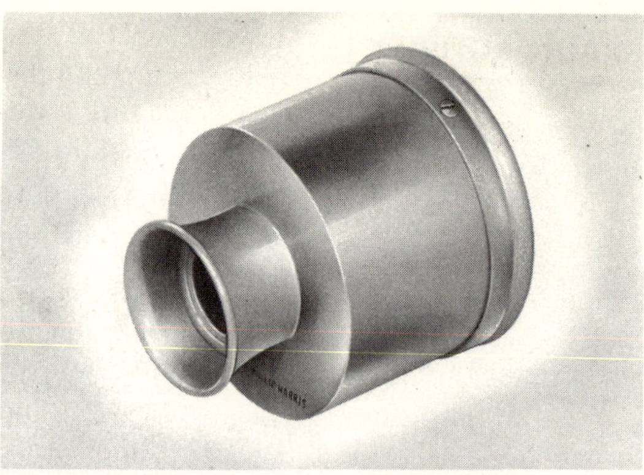

THE SPINTHARISCOPE

You may have looked through an instrument called a *spinthariscope*. It consists of a zinc sulphide screen which emits a flash of light when and where an alpha particle hits it. A weak source is fixed inside and the screen is viewed through a magnifying lens held close to the eye. Your eye has to adjust itself to the darkness, but when it has done so, you can see minute flashes of light on the screen. These are referred to as *scintillations*. The beauty of this instrument is that it shows clearly that the alpha particles from the source come out in a random way.

THE GEIGER-MÜLLER TUBE

You have also seen a Geiger tube* connected to a scaler. The scaler is a device for counting; when switched on, it will count the electric pulses received from the Geiger tube. The tube generates such a pulse whenever a radioactive particle enters it, as it also does with a gamma ray. In each case the pulse is started by ionisation produced by the radiation.

Most scalers used in schools have a mechanical register and two 'dekatron' tubes, from which the number of counts can be read off. The drawing on the left corresponds to a reading of 3458.

Two types of GM tube are often used in schools. One has a thin-end window; the other has no such window and is usually completely encased in neoprene or rubber as a protection.

Panax scaler

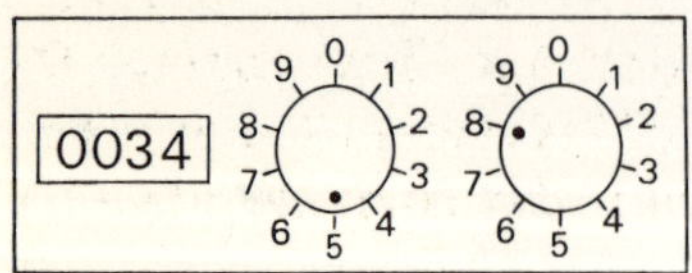

* *Sometimes called a Geiger-Müller tube or GM tube for short.*

If a radium source is held near either GM tube, the scaler records a high rate of counting. *Did you try holding a luminous watch near the GM tube? What happened?* [13] *Did anything happen when no source at all was held near the GM tube?* [14]

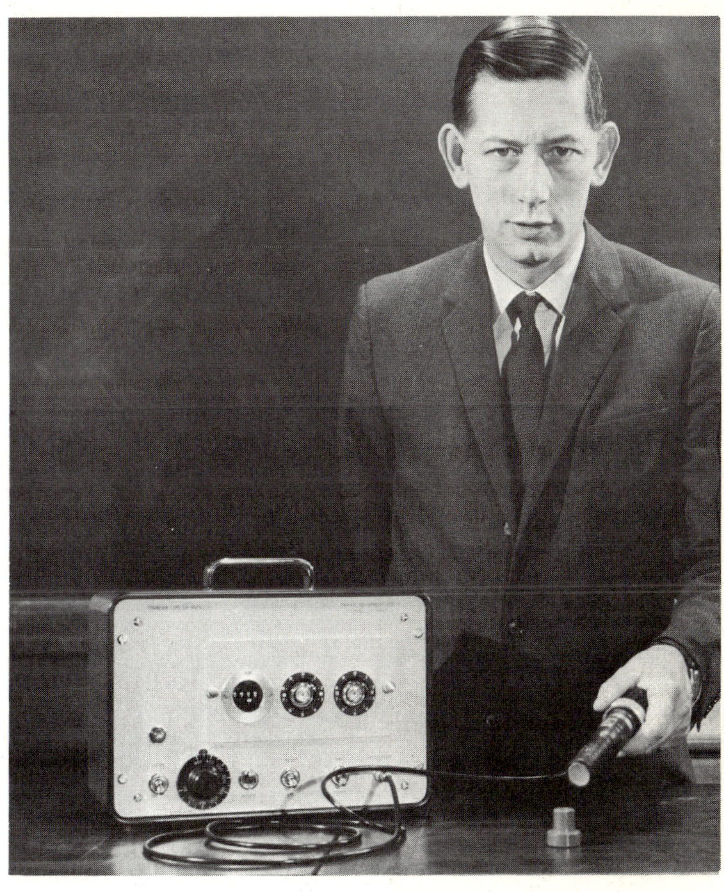

Geiger tube connected to a scaler

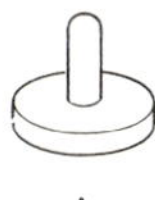

A

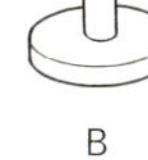

B

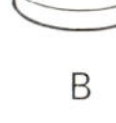

C

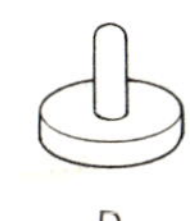

D

A DETECTIVE GAME

Suppose you have four radioactive sources, A, B, C and D. A is plutonium-239 or ^{239}Pu for short. B is strontium-90 or ^{90}Sr. C is cobalt-60 or ^{60}Co. D is radium-226 or ^{226}Ra. The significance of these names will become clear later in this book.

Let us conduct a detective game, looking for clues, in order to find out what we can about the radiation from these sources.

Source A is tried first. Its radiation discharges a gold-leaf electroscope and causes sparks on the spark counter. It has no effect at all on either GM tube connected to scalers. It is also found that the radiation is completely stopped by a piece of paper held between it and the spark counter.

Source B is tried next. It has no effect at all on the spark counter. It is held near a charged electroscope: at first it appears to have no effect, but it is eventually seen to discharge it very slowly. What can we deduce from this? One possibility is that the radiation is like that from A but very much weaker.

Source B is then tried with the thin-window GM tube and the scaler shows a high count-rate. This clue shows that it must be very different from A: it cannot be a weak version of A, or A a weak version of it. It is found that its radiation passes through paper quite easily, but it is stopped by a sheet of perspex put between the source and the GM tube. It has no effect on the thick-walled GM tube: it obviously cannot penetrate the wall.

Source C is then tried. It has no effect on either the spark counter or the electroscope. It does affect both the GM tubes. Will its radiation be stopped by perspex, as is the radiation from B? No, it gets through easily. To cut down the count-rate it is necessary to use lead.

These clues show that the radiation from source C is very different from that from source A. We also know there is something it can do which source B cannot, namely, it can affect a thick-walled GM tube. This does not, however, show that the radiation from C is different from that from B; it may be giving out the same kind of radiation, but with much higher energy. To show that B and C are different, we must find something which B can do but C cannot.

Fortunately there is such an experiment. If source B is set up on one side of a lead block with the GM tube on the other, as illustrated opposite, no radiation reaches the tube. But if a magnet is placed on top of the block, the radiation is deflected, reaches the tube and a count-rate is observed. *What would happen if the magnet were put the other way round?*[15]

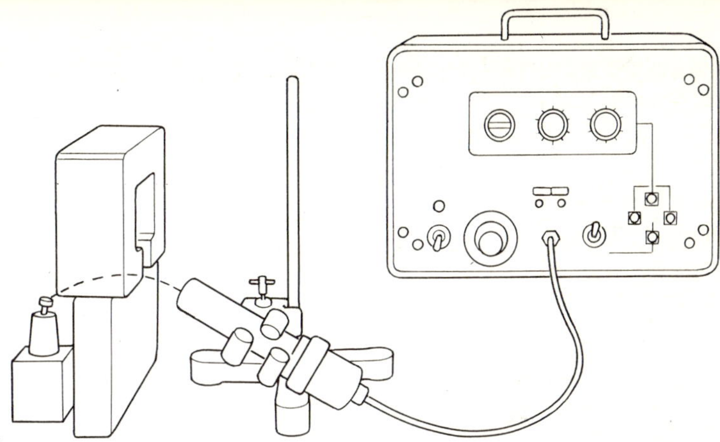

Source B Magnetic deflection ✓

Source C Magnetic deflection ✗

Source D Electroscope ✓
Spark counter ✓
Thin GM tube ✓
Thick GM tube ✓
Magnetic deflection ✓

If the experiment is repeated with source C, no deflection is observed. This shows that source B gives a different kind of radiation from source C.

These are all experiments which you can see at school. The detective work which is based on them and has been described above shows that sources A, B and C must be giving off three different kinds of radiation. That is the case. For all practical purposes in schools source A (^{239}Pu) is an alpha emitter, source B (^{90}Sr) is a beta emitter, source C (^{60}Co) is a gamma emitter.*

Source D is a radium source (^{226}Ra). Its radiation does all that source A can do, namely discharge a gold-leaf electroscope and operate the spark counter. It can do all that source C can do, namely operate both GM tubes. If it is used for the magnetic deflection experiment, it is seen that the radiation can be deflected as it can for source B. With a spark counter, the radiation from source A is stopped by paper. In the magnetic experiment, the radiation is stopped by perspex. With a GM tube, the radiation passes through paper and perspex; lead is needed to stop it. Clearly it must emit a mixture of alpha, beta and gamma radiation.

(This investigation shows that there are at least *three* types of radiation. It has not shown that there are not more. In fact, these three are the main ones from radioactive substances. However, when a beta particle is emitted, a neutral particle called a *neutrino* comes out as well, this need not concern us at present as it is not detectable with ordinary equipment.)

The cobalt-60 source does in fact emit some very weak beta radiation as well, but the covering over the source absorbs all this beta radiation.

THREE TYPES OF RADIATION

This work has confirmed what we had already suspected from cloud-chamber photographs about three types of radiation. We now have some experience of the properties of each of these radiations.

Alpha radiation

produces large numbers of ions
discharges an electroscope
produces dense tracks in a cloud chamber
operates a spark counter
has a limited range
can be stopped by paper
can be deflected in a magnetic field provided the field is very strong, and the direction of the deflection suggests positive charge

Beta radiation

produces ions, but not as many as alpha particles
discharges an electroscope, but less easily than alpha particles
produces tracks in a cloud chamber but these are less dense than alpha-particle tracks
will affect a thin-walled GM tube
can easily be deflected in a magnetic field, and the direction of the deflection suggests negative charge
passes through paper but can be stopped by perspex

Gamma radiation

produces ions, but fewer than alpha and beta radiation
does not appear to discharge an electroscope* nor operate a spark counter
travels in straight lines
cannot be deflected by a magnetic field
affects thin-walled and thick-walled GM tubes
has high penetrating power, requiring a lead block to reduce it

Although school sources of gamma radiation do not appear to discharge an electroscope, much stronger gamma sources produce enough ions to do so.

We have learnt about these properties from our experiments. In the next section we can read about the early history of radioactivity and something about the great names associated with it.

The last ten years of the nineteenth century witnessed advances in physics which were to affect the twentieth century profoundly. The study of electron streams led to the discovery of X-rays in 1895 by W. C. Röntgen, and this in turn led to the discovery of radioactivity by Henri Becquerel in 1896. In this chapter, we will be thinking about this early history.

CATHODE RAYS

It was known in the eighteenth century that a discharge of static electricity could be accompanied by a flash of light. In 1752, William Watson showed that a discharge could pass more readily through a gas at low pressure than it could at atmospheric pressure.

You have already learnt in your school course how the passage of electricity through a liquid is caused by electrically charged particles, often atoms or molecules which have acquired a positive or negative charge. These are usually called *ions*, from the Greek word meaning traveller, because they travel from one point to another when a voltage difference is applied between two electrodes. (The positive is called the *anode* and the negative the *cathode*, derived from the Greek *ana*, up, and *cata*, down.) Those ions with a positive charge travel one way, toward the cathode. Those with a negative charge travel the other way, toward the anode.

Air normally contains a few ions, and when a voltage difference is applied between two electrodes the ions will move. But if the air pressure is atmospheric, the ions will collide with the air molecules and lose their energy, and there will be no appreciable current. If the pressure is reduced, the ions will collide less often. They will therefore acquire more energy as they move through the field and any collisions may result in the formation of more ions; this is a phenomenon known as *ionisation by collision*. In this way a large current can pass at these low pressures.

Faraday made the first systematic investigation of electric discharge through gases in 1838, but vacuum pumps were not good in his day. However, in 1854 an improved

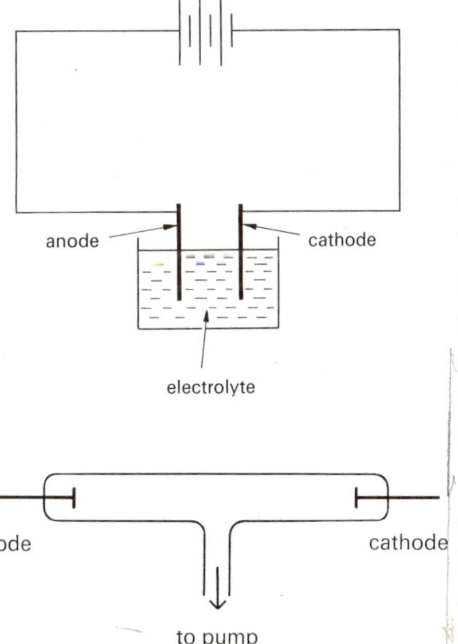

pump was developed in Germany. It also became possible to seal wires into glass tubes, so that the wires could act as electrodes. This enabled studies of electrical discharges through gases to continue. Observations showed that the glow in the tube was caused by rays from the cathode, so that in 1876 Goldstein called them *cathode rays*.

William Crookes in England, Lenard in Hungary and Perrin in France began a systematic investigation of these rays. It was shown that they travelled in straight lines (as they were able to cast a shadow of an obstacle in their path), and they could be deflected by magnetic and by electric fields. In 1895 Perrin directed the rays to fall into a cylinder connected to an electroscope and showed that they were negatively charged.

It was J. J. Thomson in 1897 at the Cavendish Labora-

J. J. Thomson (photograph copyright Cavendish Laboratory, Cambridge)

tory in Cambridge who completed this work by assuming that the cathode rays were negatively charged particles, then deflecting them in both electric and magnetic fields and using his observations to deduce both the velocity of the particles and the value of e/m, the ratio of the charge to the mass. It was this famous, classical experiment that led Thomson to conclude that the carriers of the electric charge in the cathode rays are all the same. These carriers were called *electrons*.*

In your school course you will have seen experiments very similar to those described above, except that they were not electrons produced in a discharge tube, but electrons given off by a hot filament. Those electrons travelled in straight lines, cast shadows and were deflected in electric and magnetic fields.

Below: deflection in a magnetic field

Below right: Maltese Cross experiment

* *Although the idea of electrons as particles was accepted earlier, it was not until Millikan's famous experiment in 1911, showing that all electrons had the same charge, that it was finally confirmed that electric charge was 'particulate'.*

X-RAYS

The German physicist W. C. Röntgen observed the luminescence produced when cathode rays struck the glass walls of the discharge tube.

Intense luminescence is produced if the tube is coated

Röntgen

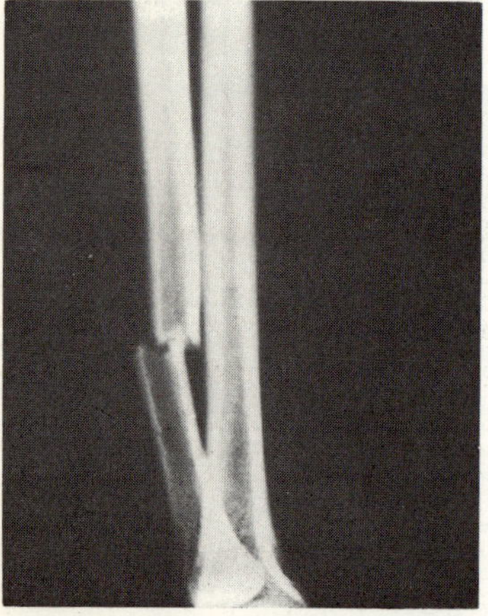

X-ray photograph of a broken bone

X-ray photograph of three-year-old child who has swallowed a plastic ring

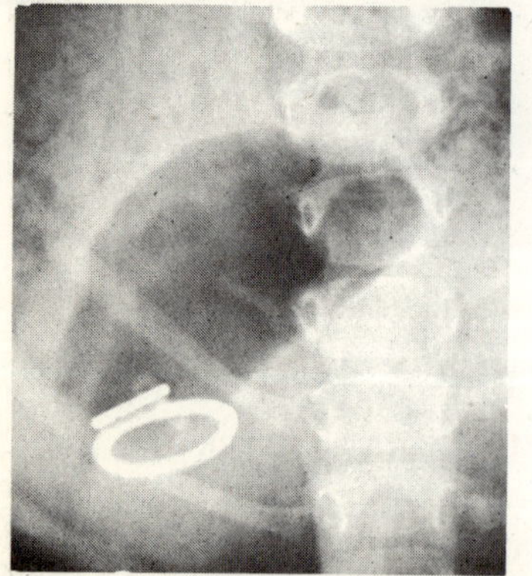

at the end with zinc sulphide and even more so with platinobarium cyanide. *What screen do you know which is given such a coating and which fluoresces when electrons hit it?* [16]

On 8 November 1895 Röntgen enclosed a discharge tube in a box of thin black cardboard in a darkened room. Near the box lay a screen of platinobarium cyanide. Röntgen noticed that the screen showed brilliant luminescence as soon as the tube was operated. It was known that cathode rays themselves could not penetrate more than a few millimetres into air at normal pressure. Clearly a new kind of radiation was involved, to which Röntgen gave the name X-rays, perhaps because their nature was unknown and X is associated with the unknown in algebra.

Within six weeks of his discovery Röntgen had a paper ready on the properties of these new rays, and within a year over sixteen papers had been written on them in various parts of the world. But it was not until 1912 that it was finally shown that X-rays are a form of electromagnetic radiation; in other words they resemble light, but have a much higher frequency and a much shorter wavelength.

Not only could the X-rays excite fluorescence, but they could darken photographic plates, even when these were wrapped in paper or enclosed in a box. This meant that substances opaque to light were transparent to X-rays, and this led Röntgen to try other opaque bodies. By exposing a hand to these rays it was possible to reveal the internal structure owing to the different degrees of transparency of different parts: flesh let through the X-rays much more easily than bone.

Röntgen found that the X-rays originated from the point where the cathode rays hit the glass. In fact X-rays are produced whenever electrons strike matter.

In early tubes the cathode was concave in shape in order to focus the cathode rays on to the target from which X-rays were given off. In a modern tube the electrons come from a hot filament and are accelerated in a strong field to hit the copper *anticathode* with an inset target of tungsten. The electrons are stopped by the target and give up their energy as X radiations, though in fact a large fraction of the energy goes as heat.

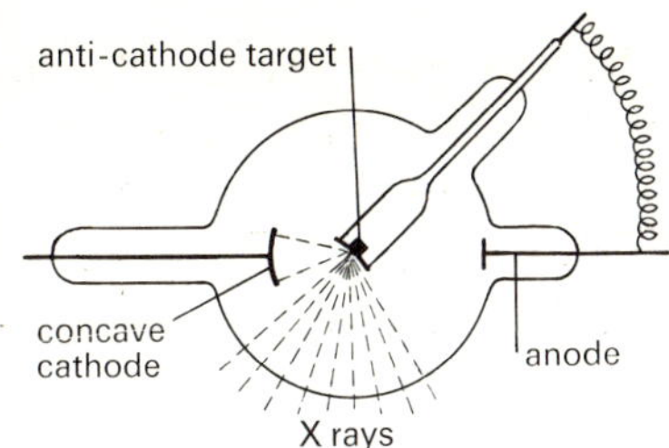

There are two basic controls on a modern X-ray tube. You can increase the current through the filament and you can increase the accelerating voltage between the filament and the anticathode. *What different effects do you think result from these controls?*[17]

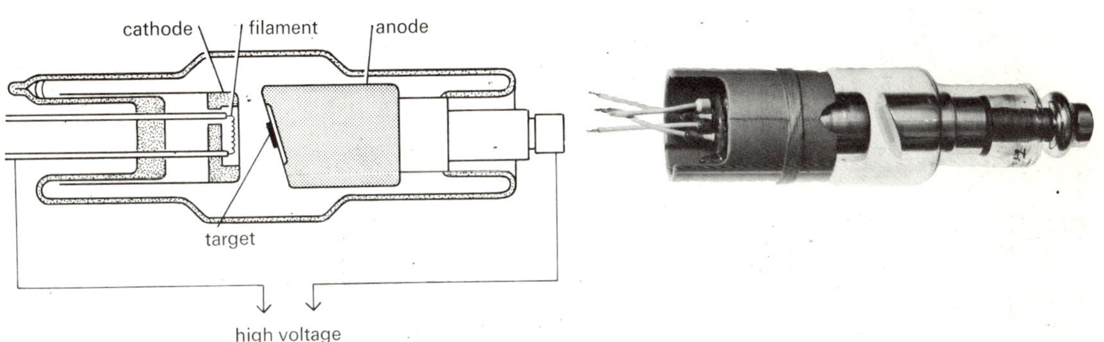

Modern X-ray tube

An important characteristic of X-rays is that they have the ability to produce ions in the gases through which they pass. They will discharge an electroscope; they will make air conduct. On the other hand, being electromagnetic radiation, they are not affected by electric or magnetic fields any more than light is.

HENRI BECQUEREL

In February 1896 the French physicist Henri Becquerel became interested. He was studying *phosphorescence*, a phenomenon in which certain substances emit weak light for a time after they have been exposed to a strong light, and he wondered if there might be some connection with the *fluorescence* which appeared on the walls of a cathode-ray tube and were known to emit X-rays. Was it possible that ordinary materials which were made phosphorescent by visible light emitted a penetrating radiation like X-rays?

Becquerel had in his possession a specimen of pure potassium uranium sulphate which was phosphorescent. He wrapped a photographic plate in paper, placed the uranium salt on top and exposed it to sunlight. When the

Becquerel

plate was developed, it was found to be darkened, showing that the uranium salt emitted a radiation which could penetrate paper. He then showed that it could penetrate thick sheets of other substances. He still thought that it was due to the exposure to sunlight.

On 26 February 1896 he prepared a photographic plate and some uranium salt, but as the sun did not shine he put them in a drawer. On 1 March there was still no sun, so he developed the plate, expecting only a faint result, but he found the plate strongly affected. In other words, the radiation from the uranium salt came from the uranium itself and was not affected by sunlight: it was nothing to do with the phosphorescence he had started to study. Thus Becquerel had discovered a new phenomenon, to which Madame Curie subsequently gave the name *radioactivity*.

Marie and Pierre Curie

MADAME CURIE

Further investigation by Becquerel showed that radiation came from other compounds of uranium and from uranium metal itself, but he was not able to make quantitative measurements. His work, however, aroused the interest of the Polish-born Marie Sklodowska Curie, then studying in Paris. With her husband, Pierre, she devised a method of measurement. She concluded that 'all the compounds of uranium studied are radioactive and, in general, the activity is greater the more uranium they contain'. The radiation was an atomic phenomenon and did not depend on its combination with other substances or on its physical state.

This led Marie and Pierre Curie to search for other substances emitting radiation, and they soon found that thorium and its compounds were also radioactive. They also found that the mineral known as pitchblende, which contains both uranium and thorium, showed much more activity than could be explained by the presence of these elements alone. Madame Curie assumed at first that there was an error in the experiment – doubt is often the scientist's first response to an unexpected phenomenon – but repeated experiments confirmed the result. Then began a

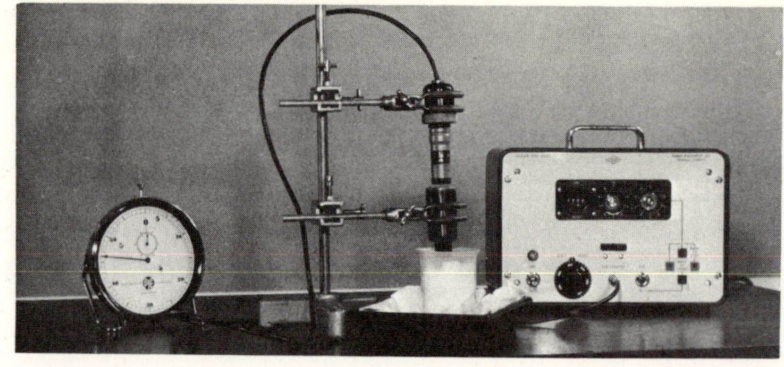

The protactinium experiment

On the left below are some figures for an experiment in which the activity of a sample of the radioactive element protactinium (^{234}Pa) was measured at frequent intervals for several minutes. The apparatus is shown in the photograph above.

Table of activity (measured in counts per second) of protactinium

Time (s)	0	10	20	30	40	50	60
Counts	0	330	610	900	1150	1380	1590
Activity (counts per second)		33	28	29	25	23	21

Time (s)	70	80	90	100	110	120	130
Counts	1760	1950	2090	2250	2370	2480	2610
Activity (counts per second)	17	19	14	16	12	11	13

Time (s)	140	150	160	170	180	190	200
Counts	2700	2790	2870	2950	3020	3080	3150
Activity (counts per second)	9	9	8	8	7	6	7

Time (s)	210	220	230	240
Counts	3210	3250	3290	3335
Activity (counts per second)	6	4	4	4

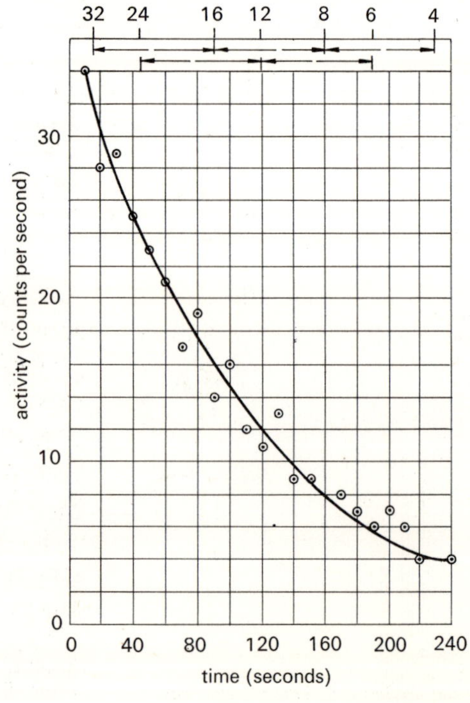

When the activity (the counts per second) is plotted against time, the graph above is obtained. The similarity with the decay curve for dice is striking.

lengthy and tedious series of chemical separations which resulted in the separation first of *polonium*, later of *radium*.

In July 1898, in the *Proceedings* of the Academy of Science, the Curies wrote:

Certain minerals containing uranium and thorium (pitch-blende, chalcolite, uranite) are very active from the point of view of the emission of Becquerel rays. In a previous communication, one of us showed that their activity was even greater than that of uranium and thorium, and stated the opinion that this effect was due to some other very active substance contained in small quantity in these minerals . . . We believe the substance we have extracted from pitchblende contains a metal not yet observed, related to bismuth by its analytical properties. If the existence of this new metal is confirmed we propose to call it *polonium*, from the name of the original country of one of us.

In 1902 they finally isolated sufficient pure radium to measure its atomic weight. They found that radium had an activity per gramme over a million times greater than the activity of uranium. Their work is a supreme example of perseverance.

C. T. R. Wilson

C. T. R. WILSON

C. T. R. Wilson described the origin of his work on cloud chambers as follows:

In September 1894 I spent a few weeks in the Observatory which then existed on the summit of Ben Nevis, the highest of the Scottish hills. The wonderful optical phenomena shown when the sun shone on the clouds surrounding the hill-top, and especially the coloured rings surrounding the sun (coronas) or surrounding the shadow cast by the hill-top or observer on mist or cloud (glories), greatly excited my interest and made me wish to imitate them in the laboratory. At the beginning of 1895 I made some experiments for this purpose – making clouds by expansion of moist air . . . Almost immediately I came across something which promised to be of more interest than the optical phenomena which I had intended to study.

You will have noticed in a warm bathroom how the invisible water vapour will condense on a cold tap or on the

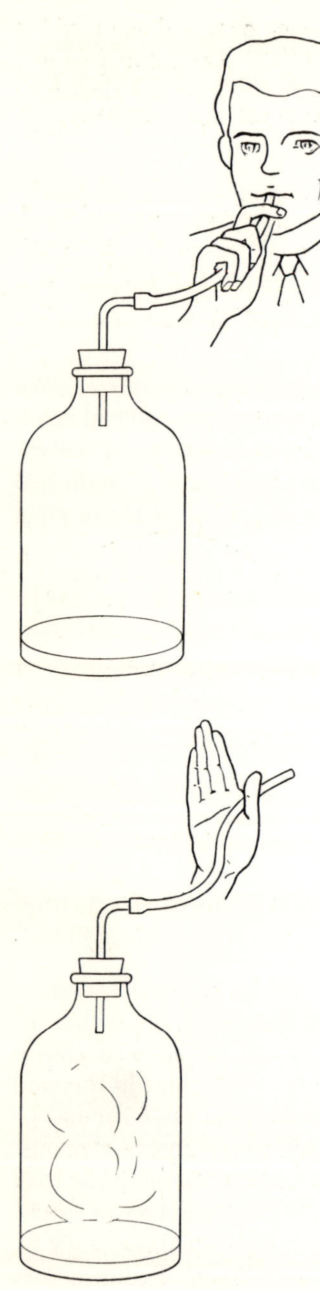

window. The air near the cold object is cooled down so that it can no longer hold the vapour present and the vapour condenses.

Air may also be cooled by causing it to expand suddenly. If you raise the pressure inside a flask containing a little water, allowing it to come to room temperature, and then cause a sudden expansion by releasing the excess air, you will see a cloud form inside.

In this experiment the drops which form the cloud are condensing on tiny dust particles. Wilson developed a chamber in which he could control the amount of expansion very precisely. He found that condensation on dust particles would occur for an expansion of about $1 \cdot 05$ times, but that if the cloud were formed several times and allowed to settle each time, a dust-free space could be obtained.

When such an expansion occurs, we say that the air has become *super-saturated* with water vapour, that is, it contains more water vapour than it can normally hold. Wilson found that if the volume increased $1 \cdot 38$ times, corresponding to an eightfold super-saturation, clouds would form in the dust-free space by condensation on the air molecules themselves. But for an expansion of $1 \cdot 25$ times, corresponding to a fourfold super-saturation, condensation would occur on any ions which were present. As alpha particles, beta particles and X-rays can all produce ions, the possible experiments which might be done with the apparatus become obvious. C. T. R. Wilson wrote:

Much time was spent in making tests of the most suitable form of expansion apparatus and in finding an efficient means of instantaneous illumination of the cloud particles for the purpose of photographing them. In the spring of 1911 tests were still incomplete, but it occurred to me one day to try whether some indication of tracks might not be made visible with the rough apparatus already constructed. The first test was made with X-rays, with little expectation of success, and in making an expansion of the proper magnitude for condensation on the ions while the air was exposed to the rays I was delighted to see the cloud chamber filled with little wisps and threads of clouds – the tracks of the electrons ejected by the action of the rays. The radium-tipped metal tongue of a spinthariscope was then placed inside the cloud chamber and the very beautiful sight of the clouds condensed along the tracks of the alpha particles was seen for the first time. The long thread-like tracks of fast beta particles were also seen when a suitable source was brought near the cloud chamber.

It was this work of C. T. R. Wilson which made possible the many beautiful cloud-chamber photographs shown in this volume, and the expansion cloud chamber which you have probably seen in your school laboratory.

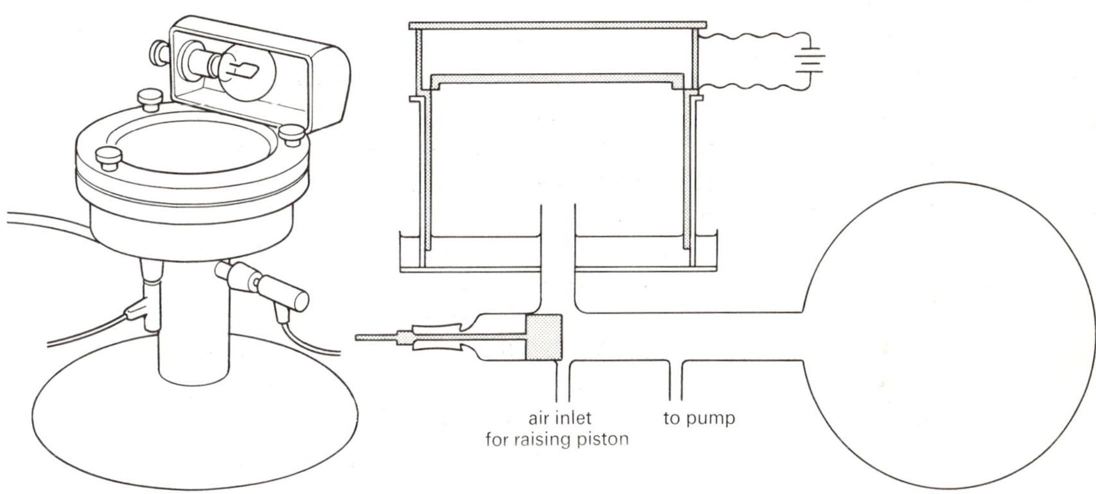

A modern expansion cloud chamber for school use

A drawing of Wilson's cloud chamber

Wilson's cloud chamber

Continuous cloud chamber

In 1936 Langsdorf suggested a diffusion cloud chamber which would be capable of continuous showing of tracks. You have seen one of these in your school course. In the type illustrated here, cooling is produced by solid carbon dioxide in the base. A felt ring in the top of the chamber is saturated with alcohol (about 2 cm³ is necessary). The vapour from this alcohol diffuses downward. As it does so, it reaches colder and colder regions so that the degree of saturation increases toward the base of the chamber. At some point it will be sufficient to cause alcohol vapour to condense on any ions formed by a radioactive source in the chamber. Illumination is necessary to make the tracks visible.

In both this chamber and the Wilson expansion chamber it is necessary to provide an electric field to clear the chamber of unwanted ions. In your school diffusion cloud chamber a suitable field can be produced by rubbing the perspex top with a woollen duster.

The bubble chamber is a more recent development. The particles pass through a tank of liquid hydrogen under pressure. The pressure is suddenly reduced and the liquid hydrogen is brought to an unstable state in which bubbles of hydrogen vapour form on ions produced along the path of any particles. Very fine tracks are obtained: a typical track is shown on the opposite page.

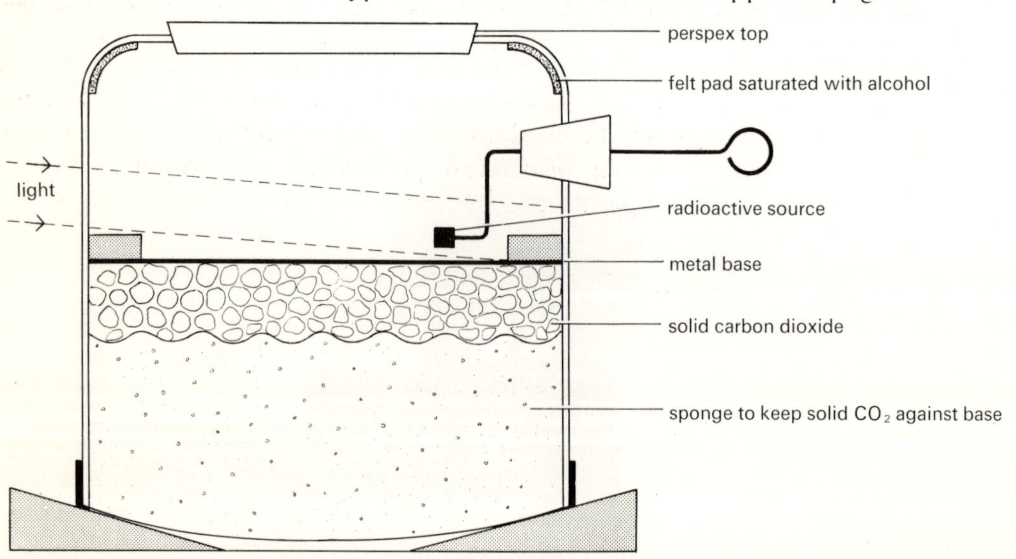

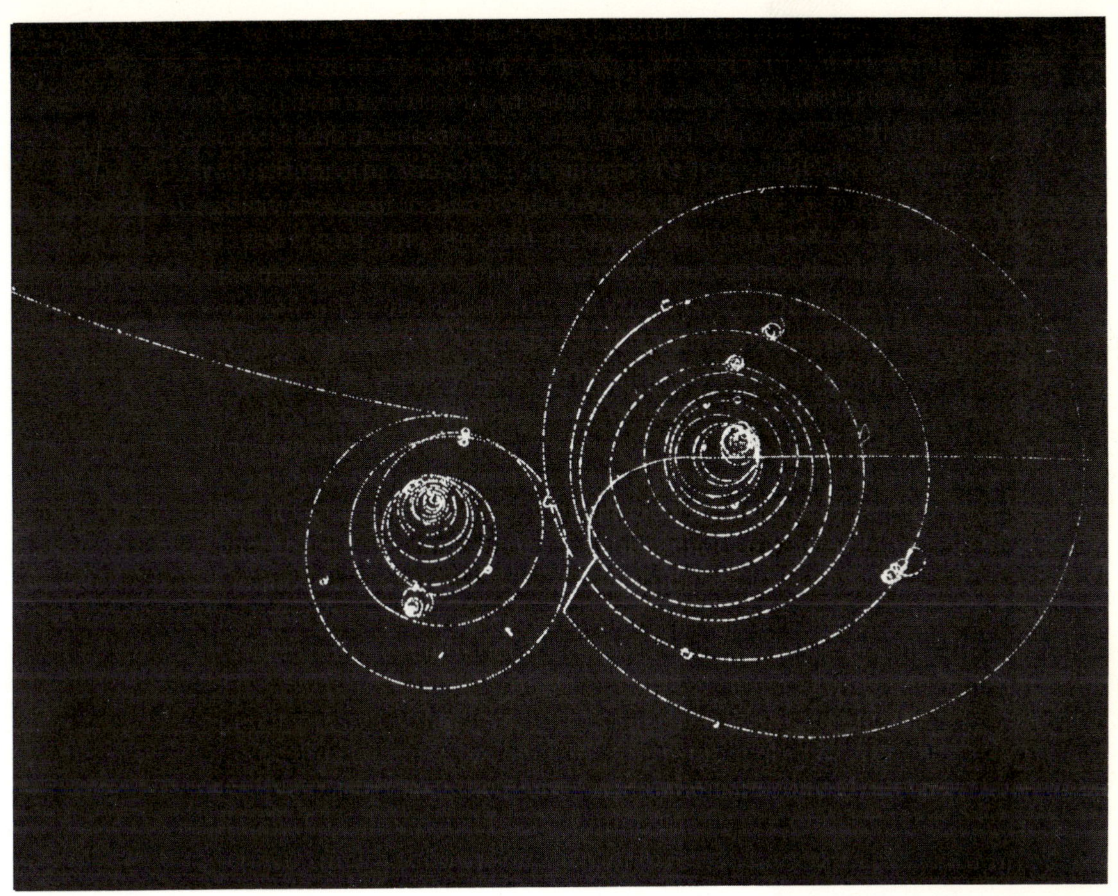

Bubble-chamber photograph in which spiralling electron tracks can be seen

This brief outline of the history of radioactivity would not be complete without reference to Lord Rutherford, but he so dominated the subject and was so much the father of nuclear physics as we know it to-day that a separate chapter must be devoted to him.

RUTHERFORD AND THE NUCLEAR MODEL

Lord Rutherford (photograph copyright Cavendish Laboratory, Cambridge)

* *To many of these substances were given special names like radium A, radium B, radium C, radium C', etc. These names are now mainly of historical interest and in this volume they will be referred to by their more usual modern nomenclature ^{218}Po, ^{214}Pb, ^{214}Bi, ^{214}Po. etc. The significance of these numbers will be explained on page 35.*

Ernest Rutherford was born in New Zealand in 1871. He studied first at Christchurch, New Zealand, and then in Cambridge, where he worked under J. J. Thomson. In 1898 he became a professor of physics at McGill University in Canada. He came to Manchester in 1907 and was made Cavendish Professor of Physics in Cambridge in 1919 in succession to J. J. Thomson. More than any other physicist he dominated the study of radioactivity and nuclear physics in the first part of this century. Apart from his own considerable contribution, his enthusiasm was infectious and he collected around him a remarkable team of physicists from all over the world.

After Rutherford's death Sir James Jeans wrote:

Those of us who were honoured by his friendship know that his greatness as a scientist was matched by his greatness as a man. We remember, and always shall remember, with affection his big, energetic, exuberant personality, the simplicity, sincerity, and transparent honesty of his character and, perhaps most of all, his genius for friendship and good comradeship.

Rutherford had the ability to draw the best out of his fellow workers, a flair for seeing the right approach to a problem and the clarity of mind to appreciate the significance of experimental results.

Thanks to his work, the distinct properties of alpha, beta and gamma radiation were fully appreciated and the nature of each was understood. He identified more radioactive substances* and understood the significance of a radioactive series in which a 'parent' radioactive substance decayed to another radioactive substance (a 'daughter product'). Perhaps his greatest achievement was his appreciation of the Geiger and Marsden experiment – see below – which led him to suggest the nuclear model of the atom. In 1919 he performed the first experiment on nuclear transmutation, in which one nucleus was bombarded by an alpha particle and turned into another. These experiments, which were the foundation of modern nuclear physics, will be discussed below.

THE NATURE OF α, β AND γ RADIATION

Through deflection experiments using very strong magnetic fields Rutherford measured e/M for alpha particles and he found a very different value from that for electrons. He suggested that the alpha particles might be helium ions – helium atoms with a positive charge (He^{++}). This was finally confirmed in an important experiment by Rutherford and Royds in 1909. This illustration of their apparatus comes from their paper in the *Philosophical Magazine* of 1909.

The experiment was made possible by blowing a very fine glass tube A. A quantity of radon – a radioactive gas, called by Rutherford *emanation*, which is the daughter product of radium – was purified and compressed into the tube A by a mercury column. This tube was sufficiently thin to allow alpha particles to pass through. The original paper reads as follows:

The thickness of the wall of the tube . . . was less than 1/100 mm and was equivalent in stopping power of the alpha particle to about 2 cm of air. Since the ranges of the alpha particles from the emanation and its products radium A and radium C are 4·3 and 4·8 and 7 cm respectively, it is seen that the great majority of the alpha particles expelled by the active matter escape through the walls of the tube A . . .

The glass tube A was surrounded by a cylindrical glass tube T, 7·5 cm long and 1·5 cm diameter, by means of a ground-glass joint C. A small vacuum-tube V was attached to the upper end of T. The outer glass tube T was exhausted by a pump through the stopcock D, and the exhaustion completed with the aid of the charcoal tube F cooled by liquid air. By means of a mercury column H attached to a reservoir, mercury was forced into the tube T until it reached the bottom of the tube A.

Part of the alpha particles which escaped through the walls of the fine tube were stopped by the outer glass tube and part by the mercury surface. If the alpha particle is a helium atom, helium would gradually diffuse . . . into the exhausted space and its presence could then be detected spectroscopically by raising the mercury and compressing the gases into the vacuum tube.

In order to avoid any possible contamination of the apparatus with helium, freshly distilled mercury and entirely new glass

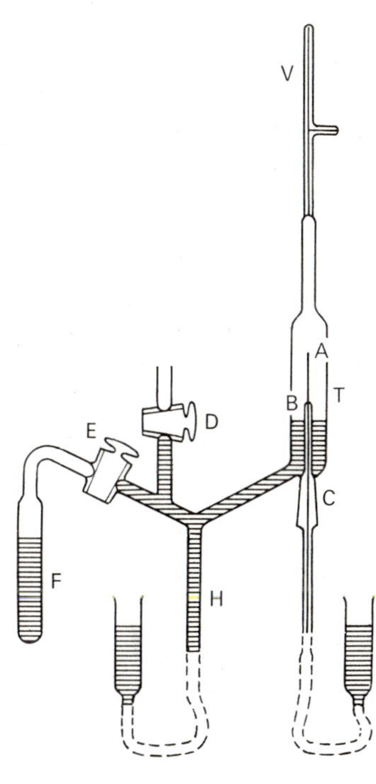

apparatus were used. Before introducing the emanation into A, the absence of helium was confirmed spectroscopically. At intervals after the introduction of the emanation the mercury was raised, and the gases in the outer tube spectroscopically examined. After twenty-four hours no trace of the helium yellow line was seen; after two days the helium yellow was faintly visible; after four days the helium yellow and green lines were bright; and after six days all the stronger lines of the helium spectrum were observed. The absence of the neon spectrum shows that the helium present was not due to a leakage of air into the apparatus.

There is, however, one possible source of error in this experiment. The helium may not be due to the alpha particles themselves, but may have *diffused* from the emanation through the thin walls of the glass tube. In order to test this point, the emanation was completely pumped out of A, and after some hours a quantity of helium, about ten times the previous volume of the emanation, was compressed into the same tube A.

The outer tube T and the vacuum-tube were removed and a fresh apparatus substituted. Observations to detect helium in the tube T were made at intervals, in the same way as before, but no trace of the helium spectrum was observed over a period of eight days.

The helium in the tube A was then pumped out and a fresh supply of emanation substituted. Results similar to the first experiment were observed. The helium yellow and green lines showed brightly after four days.

These experiments thus show conclusively that the helium could not have diffused through the glass walls, but must have been derived from the alpha particles which were fired through them. In other words, the experiments give a decisive proof that the alpha particle after losing its charge is an atom of helium.

This paper has been quoted at length partly because you may like to read the style of the actual scientific paper, and partly so that you can see the care taken to eliminate possible sources of error. Note that the experiment is repeated in order to confirm the original results.

From our subsequent knowledge of the nuclear atom we now know that a doubly ionised helium atom (He^{++}) is in fact a helium nucleus.

Deflection experiments on beta particles to measure e/m are more difficult than experiments for cathode rays or for the electrons given off by a hot filament, because beta particles do not all have the same speed when they are emitted. However, measurements were made and these confirmed that beta particles were electrons.

It took longer to confirm the nature of gamma rays, but eventually diffraction experiments with a suitable crystal were done by Rutherford and Andrade in 1914 which showed they were electromagnetic radiation; like light, but with a shorter wavelength even than X-rays.

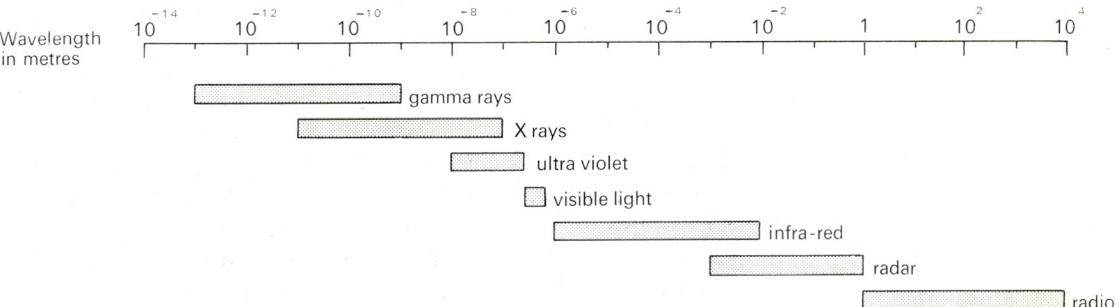

THE NUCLEAR MODEL OF THE ATOM

One of the most remarkable experiments in radioactivity, though not a spectacular one, is to set up a source of alpha particles with a detector of some sort in front of it. If some thin gold foil is put between the two, the detector reveals that the alpha particles can get through the foil.

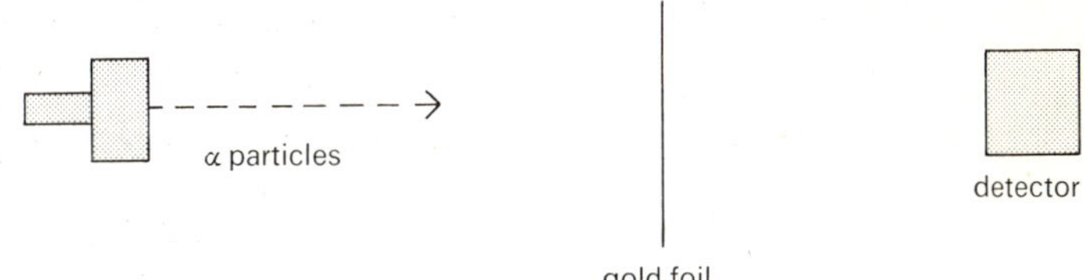

Why is this so remarkable?

The thickness of the gold might be about 10^{-6} m. A very rough estimate of the size of the gold atom is 10^{-9} m. Thus the foil might be about 1000 atoms thick. (This is only a rough estimate. It might be wrong by a factor of ten, but that would not weaken the argument. Such rough estimates are often very useful in physics.)

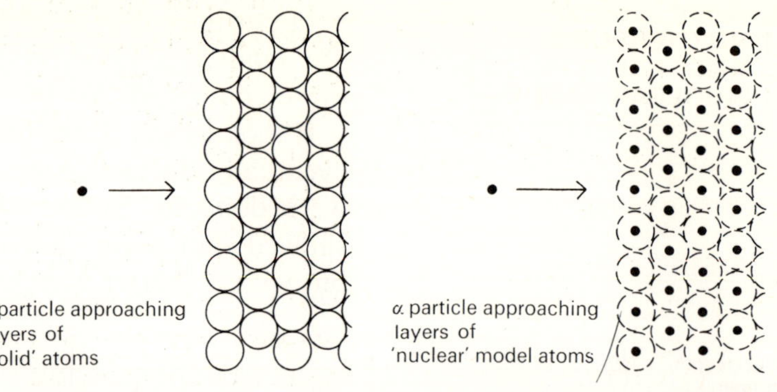

If the atoms are solid, it is incredible that the alpha particles can get through such a thickness, whatever their arrangement (or packing) may be. This led Rutherford to suggest that much of the atom must in fact be empty space as the alpha particles penetrated so easily. Confirmation came from the scattering experiment of Geiger and Marsden.

Geiger and Marsden directed fast-moving alpha particles at a gold foil and examined the scattering. They found that a small proportion was deflected through angles greater than 90° – in other words the particles appeared to be scattered *back* from the metal foil. As Rutherford said long afterward, this 'was almost as incredible as if you had fired a 15-inch shell at a piece of tissue-paper and it came back and hit you'.

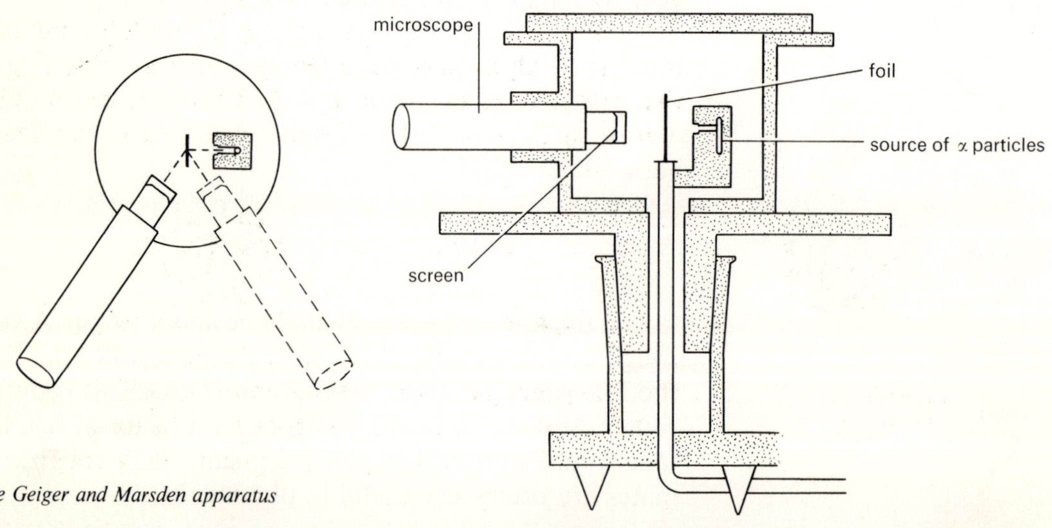

The Geiger and Marsden apparatus

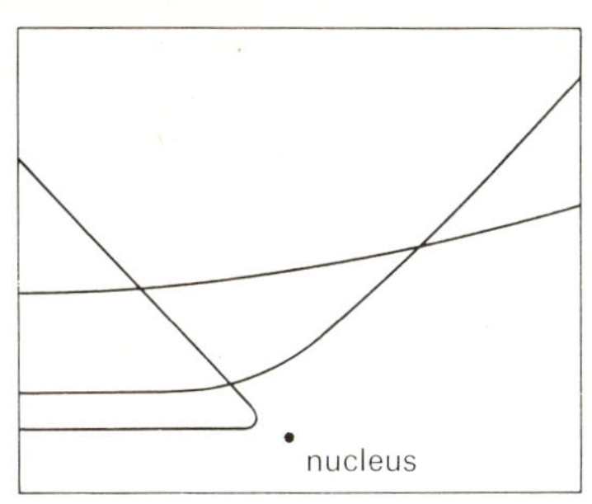

Rutherford showed that such large angle scattering could be explained only if the fast-moving alpha particles were able to move within a very close distance of the centre of the atom, very much closer than the size of the atom itself.

To explain the observations Rutherford proposed the nuclear model of the atom consisting of a central nucleus, with electrons round it. A hydrogen atom of diameter 10^{-10} m, for example, would have a nucleus 10^{-14} m in diameter. The significance of this is that the diameter of the nucleus is approximately $1/10\,000$ of the diameter of the atom, or the volume is $1/10^{12}$ the volume of the atom – and the rest is empty space.

As this book is concerned with radioactivity and not with a study of the atom, this important experiment will not be considered in greater detail here. Subsequent work suggested that the nucleus was made of protons (with unit mass and one unit of positive charge) and neutrons (with similar mass, but no charge) and that the atom as a whole consisted of the nucleus surrounded by the same number of electrons as protons so that the net charge was zero. Thus the uranium atom which has atomic number 92 and atomic mass 238 would have a nucleus of 92 protons and 146 neutrons (to bring the mass to 238) and then 92 electrons around that nucleus. As helium has atomic number 2 and atomic mass 4, its nucleus would consist of 2 protons with 2 neutrons, and 2 electrons around it.

We know that the uranium nucleus is unstable and that it emits an alpha particle (a helium nucleus, consisting of two protons and two neutrons). It will therefore turn into a new nucleus with two fewer protons and two fewer neutrons.

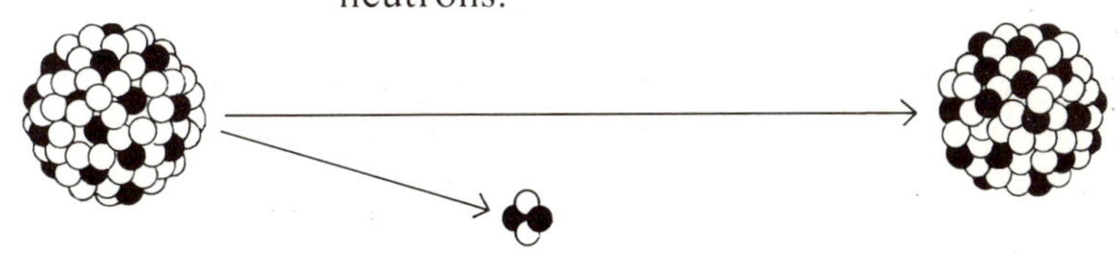

uranium nucleus $^{238}_{92}$U
with 92 protons
146 neutrons

emits α particle $^{4}_{2}$He
with 2 protons
2 neutrons

leaving thorium nucleus $^{234}_{90}$Th
with 90 protons
144 neutrons

Element	Protons	Neutrons
$^{238}_{92}$U	92	146
α ↓		
$^{234}_{90}$Th	90	144
β ↓		
$^{234}_{91}$Pa	91	143
β ↓		
$^{234}_{92}$U	92	142
α ↓		
$^{230}_{90}$Th	90	140
α ↓		
$^{226}_{88}$Ra	88	138
α ↓		
$^{222}_{86}$Rn	86	136
α ↓		
$^{218}_{84}$Po	84	134
α ↓		
$^{214}_{82}$Pb	82	132
β ↓		
$^{214}_{83}$Bi	83	131
β ↓		
$^{214}_{84}$Po	84	130
α ↓		
$^{210}_{82}$Pb	82	128
β ↓		
$^{210}_{83}$Bi	83	127
β ↓		
$^{210}_{84}$Po	84	126
α ↓		
$^{206}_{82}$Pb	82	124

We know that the beta particles emitted by a radioactive nucleus are electrons. The model of the nucleus, described above, consists only of protons and neutrons. How can we explain the emission of electrons from such a nucleus?

If a neutron were to change into a proton and an electron, the total charge would be unaltered and the mass would be approximately the same. This is the accepted explanation of beta radiation, the electron being emitted from the nucleus as the beta particle.

To give an example, the nucleus of thorium-234 consists of 90 protons and 144 neutrons. If a neutron changes to a proton and an electron, a beta particle is emitted and the new nucleus will have 91 protons and 143 neutrons. Thus $^{234}_{90}$Th changes to $^{234}_{91}$Pa when a beta particle is emitted.

RADIOACTIVE SERIES

It was in 1902 that Rutherford and Soddy first proposed a scheme of radioactive disintegration: 'The disintegration of the atom and the expulsion of a . . . charged particle leaves behind a new system lighter than before and possessing physical and chemical properties quite different from those of the original parent element. The disintegration process, once started, proceeds from stage to stage. . . .'

We now know that the uranium series disintegrates as shown on the left.

In this series, lead, for example, appears three times: $^{214}_{82}$Pb (which is radioactive), $^{210}_{82}$Pb (which is also radioactive) and $^{206}_{82}$Pb (which is stable).

These are all lead as far as chemical properties are concerned, but they have different masses. As they are all lead and have the same place in the periodic table, they are called isotopes* of lead.

A particularly interesting cloud-chamber photograph showing successive disintegration is shown opposite. It is an enlarged photograph.

* *Isotope comes from the Greek* isos, *same, and* topos, *place.*

It shows a ^{219}Rn atom decaying with the emission of an alpha particle to ^{215}Po. As the alpha particle moves off with high energy, the ^{215}Po recoils upward (and in the photograph has made intense ionisation). The ^{215}Po has then emitted an alpha particle to the right, turning it to ^{211}Pb. In the photograph this lead atom, in recoiling, has collided with an atom of the gas in the chamber and this explains the final fork.

Decay seen in a cloud chamber

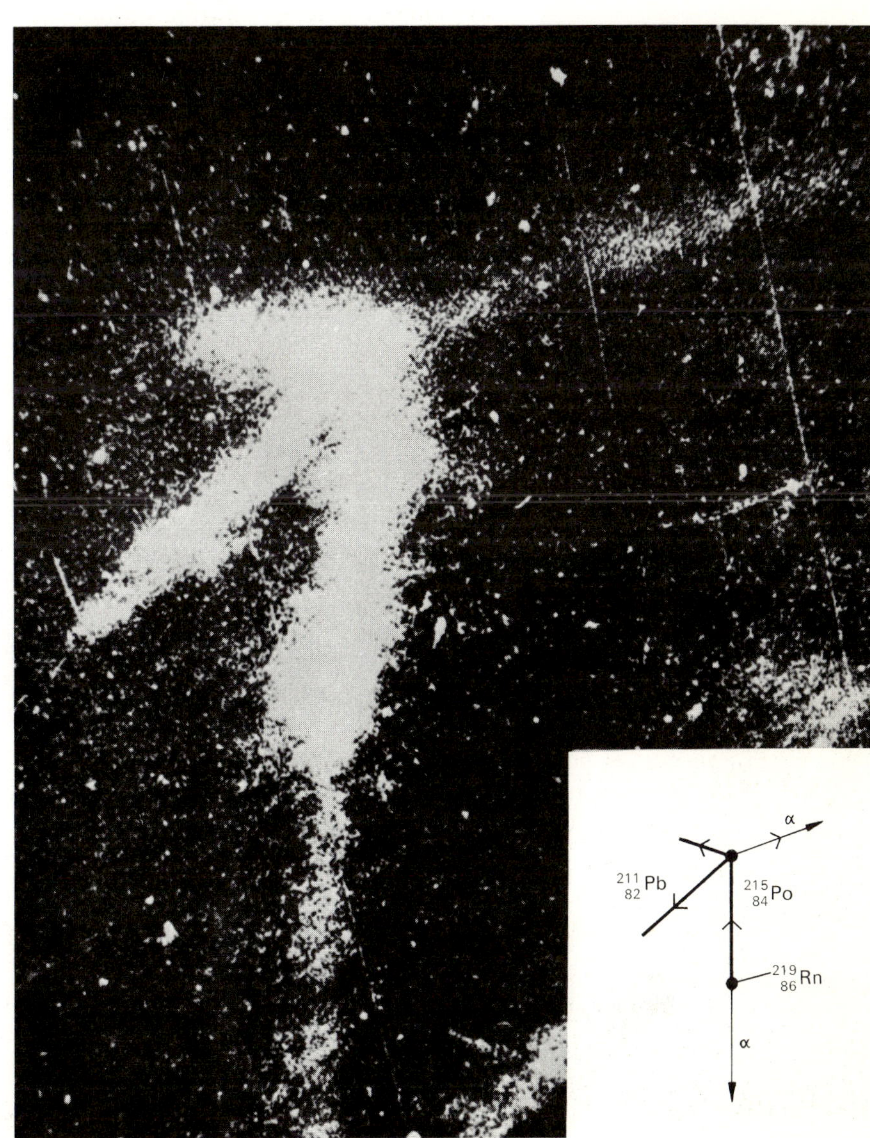

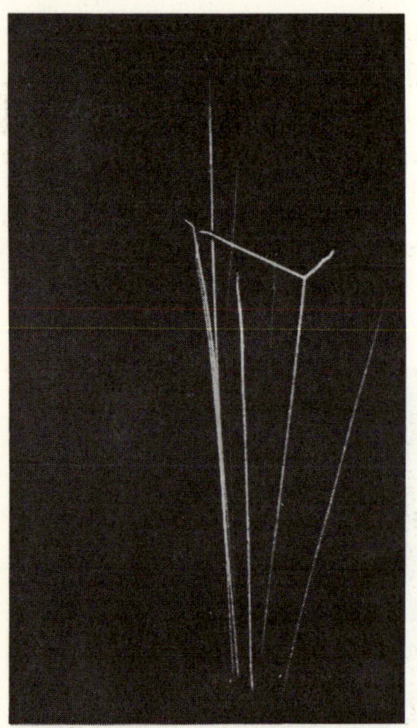

Collision with oxygen nucleus

Below: alpha particle colliding with hydrogen nucleus

Collisions with the gas in the chamber, like that mentioned above, do sometimes occur in cloud chambers. The photographs on this page show such collisions.

The photograph lower left shows the collision of an alpha particle with a nucleus of hydrogen. We can tell which is which, as the hydrogen nucleus, being lighter, produces a fainter line. The picture on the left shows a collision with an oxygen nucleus. In each case the collisions are elastic, and both momentum and energy are conserved.

The most interesting case is the photograph below in which the alpha particle has collided with a helium nucleus. Compare this with the photograph of two magnetic ring pucks of equal mass colliding together. In both cases the fork after the collision is at right angles: this occurs only when the two objects colliding have the same mass. This therefore confirms the Rutherford–Royds experiment (see page 31) identifying the alpha particle with a helium nucleus.

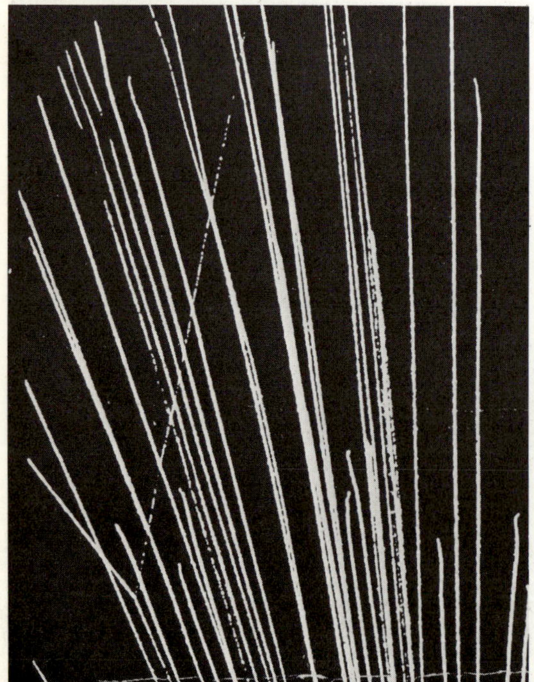

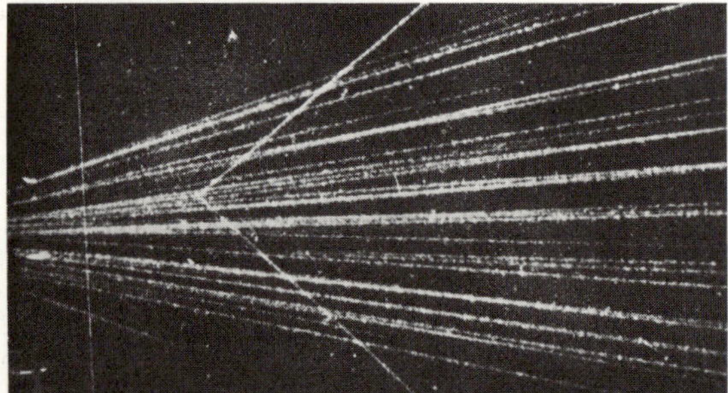

Above: alpha particles colliding with helium nucleus
Below: strobe photograph of pucks of equal mass colliding

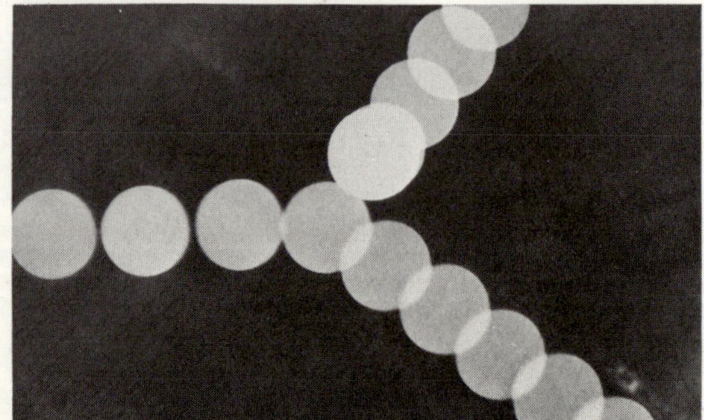

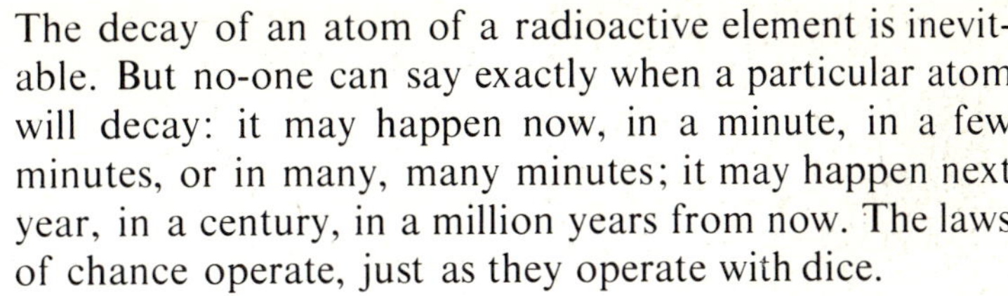

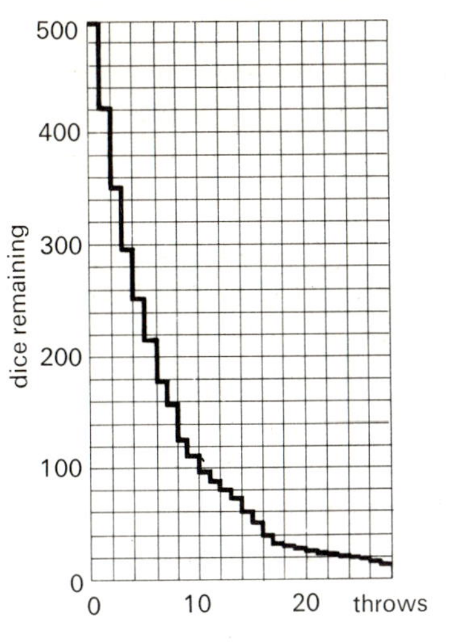

The decay of an atom of a radioactive element is inevitable. But no-one can say exactly when a particular atom will decay: it may happen now, in a minute, in a few minutes, or in many, many minutes; it may happen next year, in a century, in a million years from now. The laws of chance operate, just as they operate with dice.

Here are some figures obtained during a game in which 500 dice were thrown together and all the sixes were removed after each throw.

Throw	0	1	2	3	4	5	6	7	8	9	10
Number of sixes thrown		80	71	55	42	36	37	23	31	13	15
Number of dice left	500	420	349	294	252	216	179	156	125	112	97

Throw	11	12	13	14	15	16	17	18	19	20
Number of sixes thrown	10	8	9	11	10	10	5	2	4	3
Number of dice left	87	79	70	59	49	39	34	32	28	25

Throw	21	22	23	24	25	26	27	28	29	30
Number of sixes thrown	2	1	2	1	3	1	2	1	1	2
Number of dice left	23	22	20	19	16	15	13	12	11	9

The histogram on the left shows how the number of dice diminishes. The intriguing thing about this decay of dice' is that about four throws are needed to reduce the number of dice by one half, however many dice there are.

From 500 to 250 dice took 4 throws
From 400 to 200 dice took 4 throws
From 300 to 150 dice took 4 throws
From 200 to 100 dice took between 4 and 5 throws
From 100 to 50 dice took 5 throws
From 50 to 25 dice took 5 throws

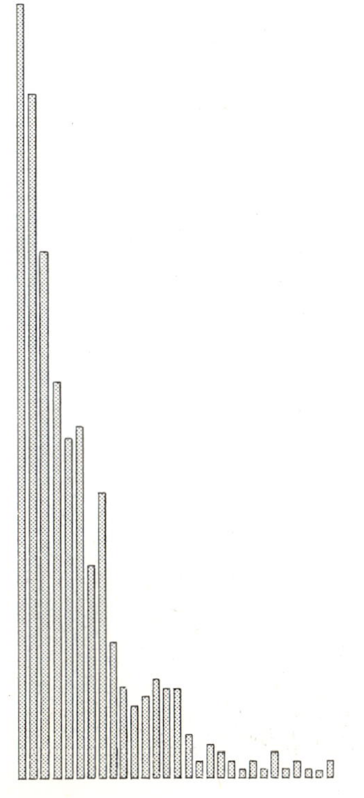

If the dice rejected at each throw are counted, a similar histogram is obtained. On the left is a drawing showing the dice which were removed at each throw piled up on end.

Each column in this histogram represents the rate at which the dice 'decay' – and this also falls to one half in about four throws.

From 80 dice per throw to 40 dice per throw took $3\frac{1}{2}$ throws
From 40 dice per throw to 20 dice per throw took 4 throws
From 20 dice per throw to 10 dice per throw took 4 throws
From 10 dice per throw to 5 dice per throw took $4\frac{1}{2}$ throws

The rate of decay (the 'activity') appears to be related to the number of dice left.

From this decay graph for protactinium we see that the activity fell

From 32 counts/second to 16 counts/second in 74 seconds
From 24 counts/second to 12 counts/second in 72 seconds
From 12 counts/second to 6 counts/second in 69 seconds
From 16 counts/second to 8 counts/second in 70 seconds

This gives an average time for the activity to reduce to one-half of 71 seconds. This time is called the *half-life.* It will also be the time that it takes for half the atoms in our sample to decay, although no-one can say which individual atoms will disintegrate.

The half-lives of naturally occurring radioactive substances vary considerably. For some elements, the half-life is a mere fraction of a second; for others it may be millions of years. Here are the values for the uranium-238 series.

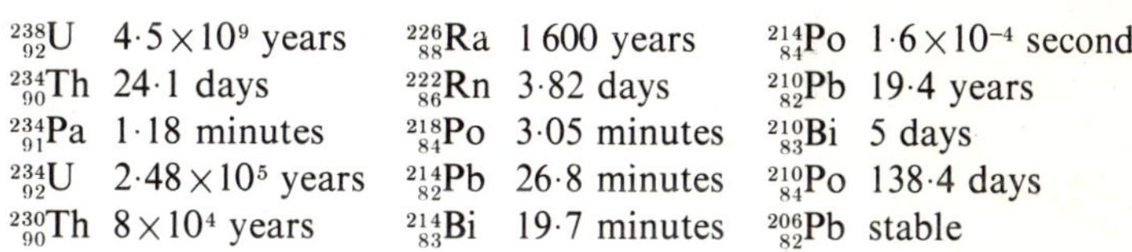

$^{238}_{92}$U $\quad 4\cdot5\times10^9$ years	$^{226}_{88}$Ra $\quad 1\,600$ years	$^{214}_{84}$Po $\quad 1\cdot6\times10^{-4}$ seconds
$^{234}_{90}$Th $\quad 24\cdot1$ days	$^{222}_{86}$Rn $\quad 3\cdot82$ days	$^{210}_{82}$Pb $\quad 19\cdot4$ years
$^{234}_{91}$Pa $\quad 1\cdot18$ minutes	$^{218}_{84}$Po $\quad 3\cdot05$ minutes	$^{210}_{83}$Bi $\quad 5$ days
$^{234}_{92}$U $\quad 2\cdot48\times10^5$ years	$^{214}_{82}$Pb $\quad 26\cdot8$ minutes	$^{210}_{84}$Po $\quad 138\cdot4$ days
$^{230}_{90}$Th $\quad 8\times10^4$ years	$^{214}_{83}$Bi $\quad 19\cdot7$ minutes	$^{206}_{82}$Pb $\quad$ stable

A very graphic way of showing this decay of a radioactive element is by using the radioactive gas, thoron (^{220}Rn), which has a short half-life of 54 seconds. In a closed bottle which contains thorium hydroxide (a powder), there will be a store of this gas: in fact, the thoron will be *in equilibrium* with the parent element thorium. This means that the rate at which the thorium is decaying to produce thoron is exactly equal to the rate at which the thoron itself is decaying. So the quantity of thoron present remains approximately constant.[*]

This thoron can be squeezed out into a diffusion cloud chamber and the decay process can be watched. On the next page is a series of photographs taken at 54-second intervals of such an experiment. The number of tracks in each picture diminishes: check whether the figure of 54 seconds is a reasonable one for the half-life.

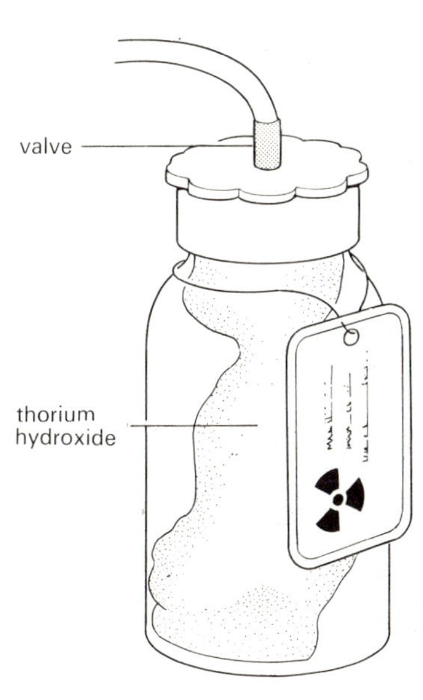

valve

thorium
hydroxide

[*] *As the parent element, thorium, is diminishing, the quantity of thoron will in fact diminish also with the same half-life, but as the half-life of thorium (^{232}Th) is $1\cdot41\times10^{10}$ years this will not be appreciable.*

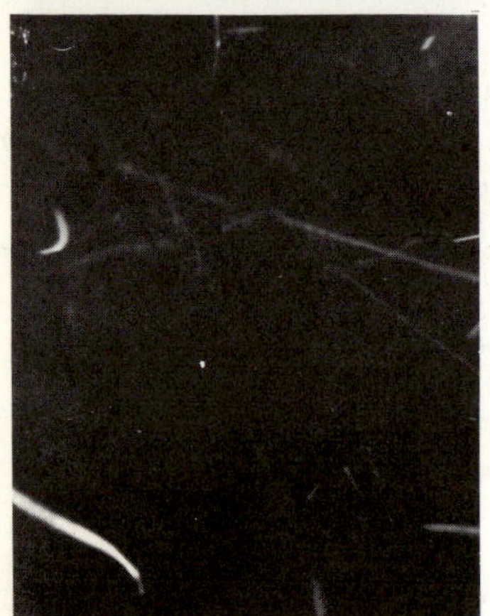

Thoron in a cloud chamber

A more careful measurement of the half-life of thoron is being made in the photograph below. The electric circuit consists of a high-voltage power supply (giving about 3000 volts), an ionisation chamber and a device to measure very small currents. Despite the high voltage, no current flows because of the air gap in the ionisation chamber. But if thoron gas is pumped into the ionisation chamber, the air is ionised and a current flows. This small current is measured, using the d.c. amplifier. As the thoron decays, so the current gets less and less. By drawing a graph of the current against time, the half-life of thoron can be found.

Experiment to measure the half-life of thoron

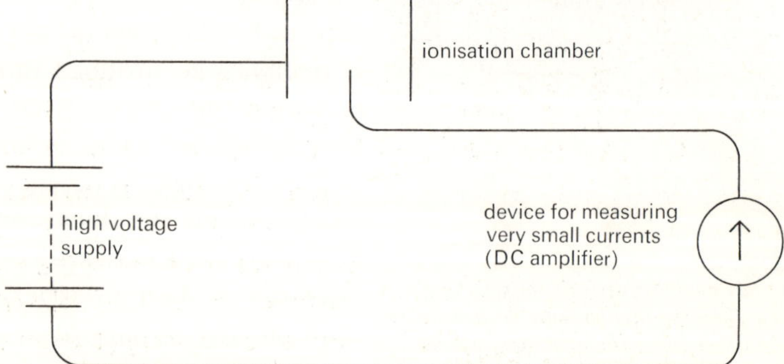

STATISTICS OF COUNTING

Radioactive decay is governed by the laws of chance and the overall picture is much more predictable for radioactive atoms than it is for dice. We had only 500 dice, but the protactinium experiment discussed on page 40 involved millions of atoms. To predict events which are controlled by chance, we do need very large numbers. Radioactive measurements with our counting equipment are the more predictable the larger the count-rates.

On the left are some figures for a series of counts, 200 in all, made with a GM tube placed at a fixed distance from a radium source. The counts in each case were for the same time interval (10^{-1} seconds).

The range of the counts is very wide – from 67 to 265. Let us add up the number of counts which occur between 60 and 69, then between 70 and 79, then between 80 and 89, and so on. We get the following table:

Range of counts	Number of occurrences	Range of counts	Number of occurrences
60–69	1	170–179	8
70–79	1	180–189	16
80–89	8	190–199	13
90–99	13	200–209	5
100–109	9	210–219	9
110–119	14	220–229	7
120–129	21	230–239	5
130–139	19	240–249	1
140–149	17	250–259	1
150–159	19	260–269	2
160–169	17		

When plotted, these counts give the first histogram on page 45.

With such a wide spread, the mean count-rate (150) is not a very reliable prediction of what the next reading will be when the experiment is repeated. We repeated the experiment with the same radium source at exactly the same distance from the GM tube, but counted in each case for ten times as long (1 second). The counts were as shown on the left.

163 158 225 186 200 196
193 145 132 168 134 153
191 190 149 148 222 161
169 177 120 163 159 139
131 103 115 137 126 124
110 140 121 133 086 127
162 219 111 082 111 123
165 205 115 151 142 097
136 195 150 093 173 161
229 227 211 143 203 188
230 192 084 195 230 217
196 092 144 188 174 129
143 141 147 181 110 152
179 074 164 162 189 223
237 190 243 178 265 190
126 168 147 215 227 195
189 155 214 262 185 128
138 116 111 158 135 161
121 214 227 157 173 111
103 181 086 233 186 123
152 142 127 156 153 124
160 126 132 131 095 095
135 114 098 130 106 109
176 139 153 113 133 123
196 113 109 131 092 111
204 164 067 090 108 120
171 151 154 092 108 186
167 210 230 252 191 169
143 141 085 086 152 151
114 158 206 155 121 123
211 093 215 090 147 160
136 128 121 096 149 107
087 088 100 143 130 131
122 095

1290 1365 1277 1260 1235
1212 1165 1215 1323 1230
. . . .

As in the previous experiment there were 200 counts. On finding the number in each range, we got the following table:

Range of counts	Number of occurrences	Range of counts	Number of occurrences
1100–1199	11	1400–1499	29
1200–1299.	88	1500–1599	7
1300–1399	61	1600–1699	4

Plotted, these counts give the second histogram opposite. The graph for these counts shows a much more pronounced peak and the spread about that peak is much less significant.

It follows that an experimenter in this field wanting precision must have large numbers of counts. Statistical theory shows that, when the number of counts is N, $\sqrt{N}$ is a measure of the spread of the counts about the average. This is called the *standard deviation.**

In practice we need to make 10 000 counts if we want a standard deviation as low as 1% ($\sqrt{10\,000} = 100$, which is 1% of 10 000). If we are satisfied with 100 counts, the standard deviation is as much as 10% ($\sqrt{100} = 10$, which is 10% of 100).

It is important to remember these statistical fluctuations when making measurements in radioactivity. Relying on only one reading can give strange results. For example, the background count measured with a GM tube and scaler might have a mean value of about 50 counts/minute, but a single count might suggest a count-rate very much higher than this.

* More precisely, two-thirds of the readings lie within one standard deviation from the mean. For a mean count rate of 100, the standard deviation is 10, and this means that two-thirds of the readings will lie between 90 and 110. You can have 95% confidence that a reading will lie within two standard deviations from the mean.

Histogram showing the spread when the number of counts is low

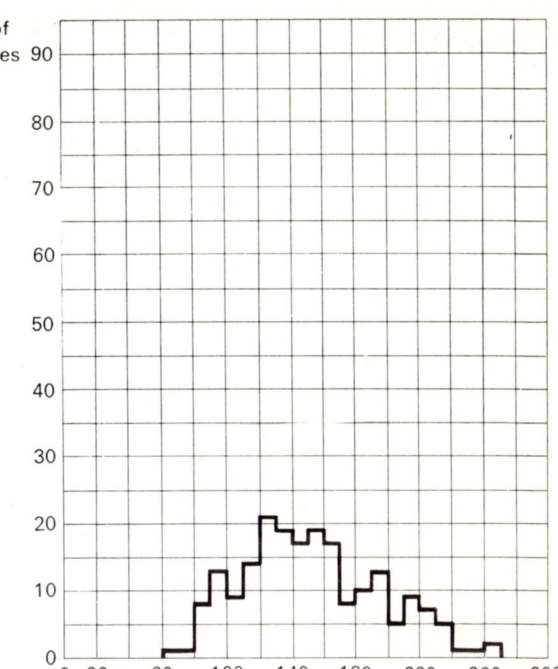

Histogram showing the spread when the number of counts is high

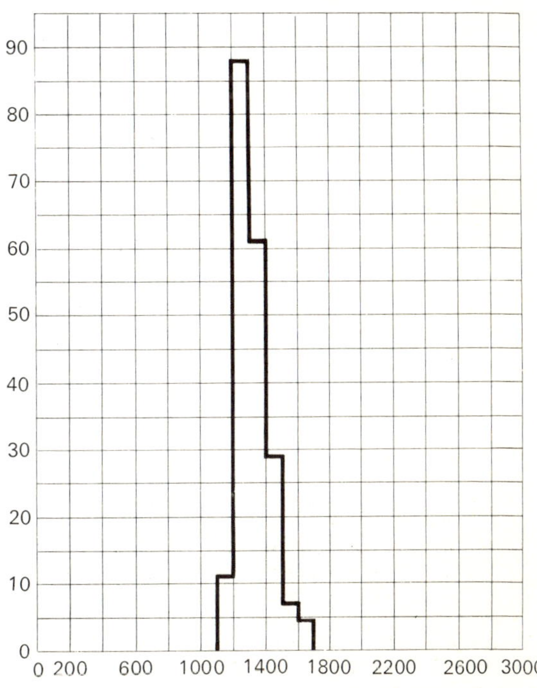

ARTIFICIAL TRANSMUTATION

The discovery of radioactive decay had shown how one element could spontaneously turn into another. In 1919 Rutherford turned one element into another artificially: a process now known as *transmutation*.

The collisions illustrated on page 38 were all elastic collisions; for example, the alpha particle hit a nitrogen nucleus and they both rebounded. Rutherford's experiment was not an elastic collision; in it an alpha particle from radium collided with a nitrogen nucleus. They came momentarily together and then a hydrogen nucleus (a proton) was given off. The reaction could be written like this

$$^{14}_{7}N + {}^{4}_{2}He \longrightarrow {}^{1}_{1}H + {}^{17}_{8}O$$

It should be noted that only about 1 in 500 000 of the alpha particles in the experiment caused transmutation, but it was typical of Rutherford's genius to note that this was happening. Nitrogen had been turned into oxygen.

The cloud-chamber photograph below shows this happening. An alpha particle has hit a nitrogen nucleus. The long track going off to the left is the path of a proton and the much heavier oxygen nucleus made the short track going off to the right of the fork.

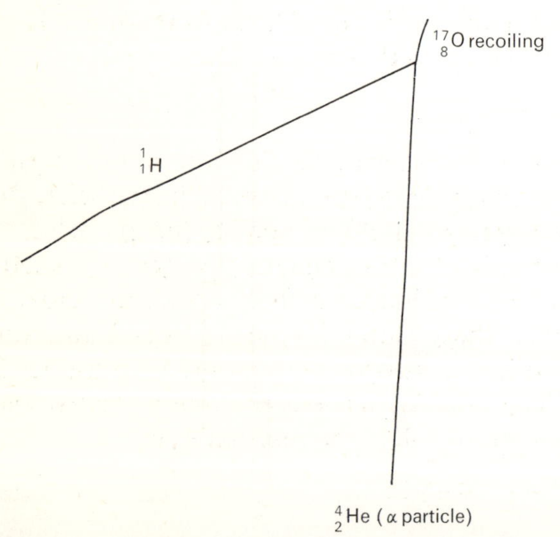

The next nuclear reaction, by Cockcroft and Walton in 1932, was made possible when a large accelerating machine was made. Lithium was bombarded with high-energy protons with the following result:

$$^1_1\text{H} + ^7_3\text{Li} \longrightarrow ^4_2\text{He} + ^4_2\text{He}$$

This is interesting because the products of the reaction are both alpha particles. If momentum is to be conserved, they must go off in opposite directions. This was found to be the case: see the cloud chamber photograph below left.

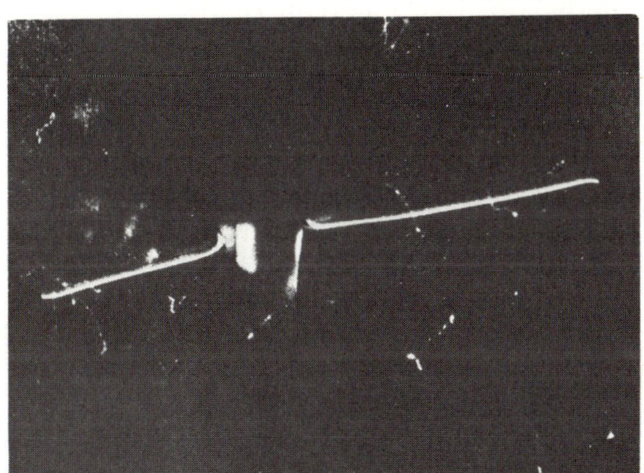 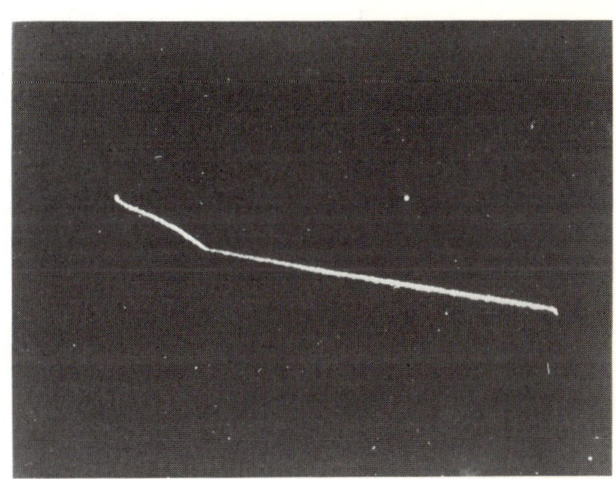

THE NEUTRON

In 1932 Chadwick was the first to confirm the existence of the neutron (the possibility of its existence had been suggested much earlier in 1920 by Rutherford). He bombarded beryllium with alpha particles:

$$^4_2\text{He} + ^9_4\text{Be} \longrightarrow ^{12}_6\text{C} + ^1_0\text{n}$$

As neutrons are uncharged, they are not repelled by a positively charged nucleus, as are alpha particles, so they become extremely useful as bombarding particles.

An example of a transmutation caused by a neutron is illustrated in the cloud-chamber photograph above right. A neutron, whose path is invisible, comes from below the picture. It hits a nitrogen nucleus and is momentarily absorbed. The resulting nucleus emits an alpha particle and the remainder recoils. The reaction was:

$$^1_0\text{n} + ^{14}_7\text{N} \longrightarrow ^{13}_5\text{B} + ^4_2\text{He}$$

THE POSITRON

In 1932 Anderson, studying cosmic rays in the United States, observed that in addition to the deflection of electrons in a magnetic field there were sometimes tracks going the opposite way. If in the diagram on the left the path ab is made by an electron, what may be the cause of the track cd? It may be a positive particle moving from c to d, but it may equally be a negative particle travelling from d to c.

It was the photograph on the left that finally confirmed the existence of the positive electron or the *positron*. In this cloud chamber there was a 6-mm lead plate across the middle. As the curvature is greater below the plate than above it, the particle could only have been moving from the top to the bottom, having lost energy on the way through the lead – and the curvature was in a direction which made it clear it was a positive particle. Thus the existence of a positive electron was established. It has the same mass as the electron, but a positive charge instead of a negative one.

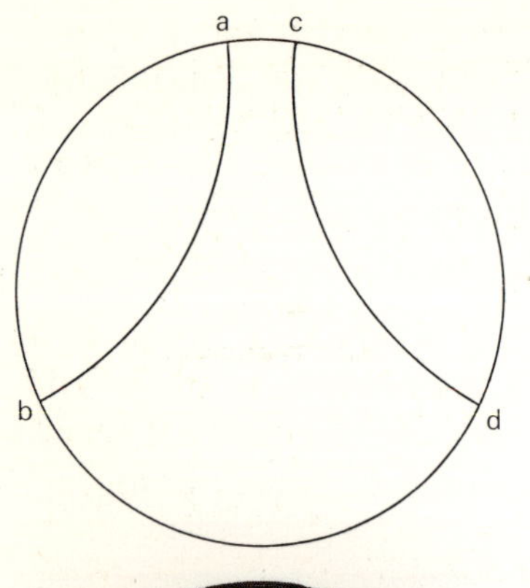

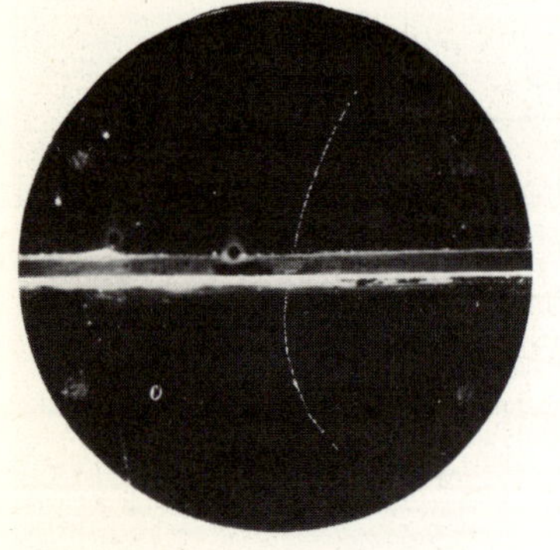

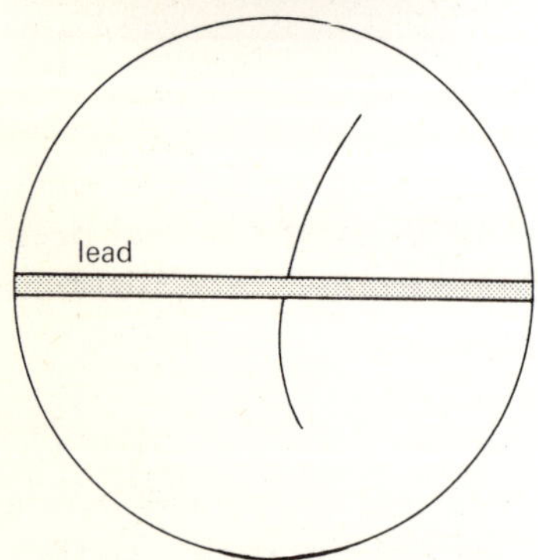

The discovery of the positron

ARTIFICIAL RADIOACTIVITY

Artificial radioactivity was discovered in 1934 by Irene and Frédéric Joliot-Curie.* They found that when aluminium was bombarded with alpha particles, positrons were emitted and that they continued to be emitted after the source of alpha particles was removed. The intensity of the positrons decayed with a half-life of about three minutes. The reaction was as follows:

$$^{27}_{13}\mathrm{Al} + ^{4}_{2}\mathrm{He} \longrightarrow ^{30}_{15}\mathrm{P} + ^{1}_{0}\mathrm{n}$$

and $^{30}_{15}\mathrm{P}$ decayed to $^{30}_{14}\mathrm{Si}$ with the emission of a positron.

Many more artificially radioactive substances were discovered. By the use of particles from accelerating machines and neutrons (which are particularly convenient as bombarding particles as they have no charge), there are now over 1000 radioactive isotopes (or *radioisotopes*) known. Some emit negative beta particles, some positive beta particles.

** Irene was the daughter of Madame Curie and she married the French physicist Frédéric Joliot. They took the name Joliot-Curie.*

Because of the large number of neutrons in a nuclear reactor, it is relatively easy to produce radioisotopes in quantity, and they have proved of great value to industry and research, as is discussed in the next chapter.

FISSION

Since this is a book on radioactivity and not on nuclear energy, only brief reference will be made to fission.

If $^{235}_{92}U$ is bombarded by neutrons, preferably 'slow' neutrons, it may break into two fragments and this is accompanied by the release of considerable energy. Such a process is known as *fission*. In addition to the two main fragments, several neutrons are also emitted; these can go on to produce further fissions and so a *chain reaction* can be set up.

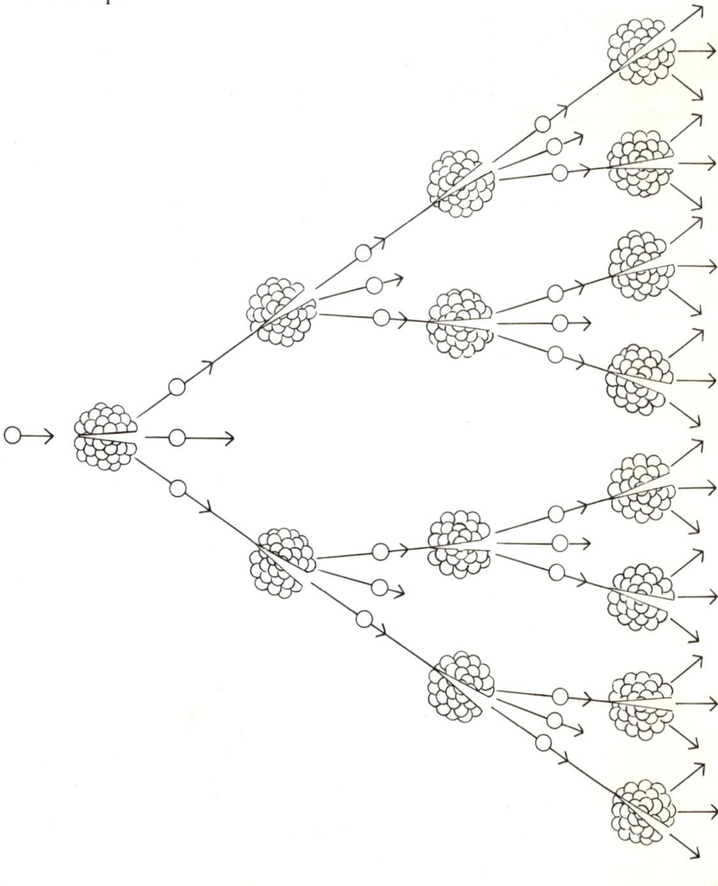

Fission tracks in a cloud chamber

The nucleus may break up in many different ways. Usually the two main fragments are unequal. Most of the fission products are highly radioactive. It is these radioactive products that remain a hazard after a nuclear explosion has taken place.

The cloud-chamber photograph above shows fission occurring. A thin layer of uranium is in the centre of the chamber. It is exposed to strong neutron bombardment. Many recoil protons and other nuclei are visible, but one example of fission is also seen. The tracks of the fission products leave the foil in opposite directions. They show characteristic branches, which increase in quantity toward the end. These branches are due to collisions with nuclei of the gas in the chamber. As the fission products are massive and highly charged, such collisions are frequent.

Radioactive isotopes have contributed much to the benefit of man in the last twenty-five years. They have aided the diagnosis and treatment of disease; they have enabled industry to produce better and cheaper products; they have contributed to agriculture and to food technology; they have made possible considerable advances in scientific knowledge.

It should be emphasised that there is no chemical difference between an element and its radioactive isotope. If a plant normally takes up phosphorus from its surroundings, it will just as easily take up radioactive phosphorus. The picture below shows part of a tomato plant. The leaf was dipped into a solution containing the radioactive isotope phosphorus-32 (^{32}P). The leaf took up the radioactive material. It was then placed on a photographic plate. ^{32}P emitted beta particles, which affected the plate, producing this picture when it was developed.

An autoradiograph of a leaf from a tomato plant

This illustrates one use of radioisotopes as *tracers*: they provide us with labelled atoms whose subsequent whereabouts can be traced photographically or with a Geiger counter or other detecting device. Their second and equally important usefulness is in the radiation which they themselves emit. Both radium and X-ray tubes have been used as sources of radiation, but radioisotopes have many advantages which will be discussed below.

The uses to which radioisotopes are put at the present time are so varied that in this volume we can only mention a small selection, but it is hoped that this will be sufficient to show what great benefits they bring.

USES OF RADIOISOTOPES IN MEDICINE

When a clot, or *thrombosis*, forms in the bloodstream, it is important to find where the blockage is. By injecting the radioisotope sodium-24 into the body, the rate of flow can be measured. It should take a certain time for the radioactive substance injected into the arm to reach the foot, and if it does not, the stoppage on the way can be traced.

The thyroid gland normally takes up iodine which enters the body. By injecting radioactive iodine (^{131}I), we can test how effectively the thyroid is working. Similarly, radioactive phosphorus can indicate the extent of brain tumours.

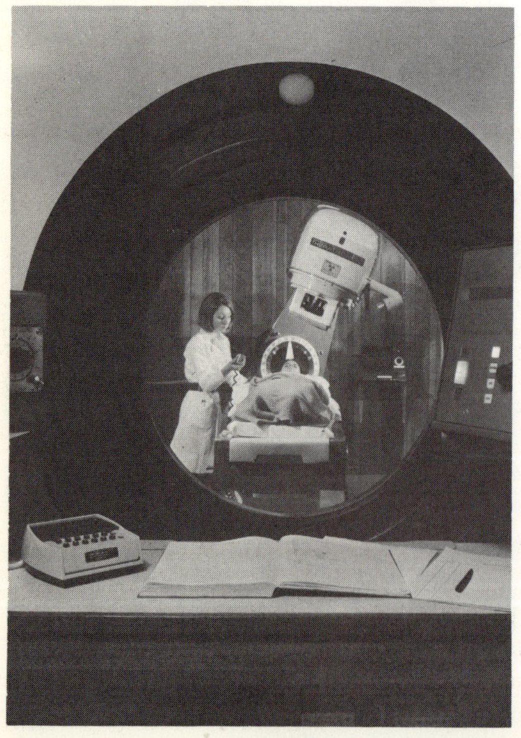

The peaceful use of radioactivity: cobalt therapy unit viewed through safety window

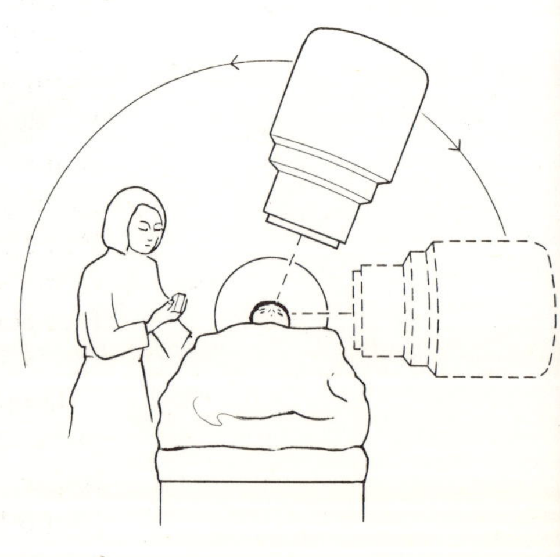

The radiation from radium has long been known in the treatment for cancer. The radiation can attack and kill malignant cells. However, radium is very expensive, and various radioisotopes such as cobalt-60 can be used instead. The photograph opposite shows a patient undergoing treatment in a cobalt therapy unit.

Unfortunately, the radiation not only kills harmful cells but can also damage healthy living material. Particular care must therefore be taken in the application of the radiation. By rotating the source it is possible to ensure that surrounding areas receive only the minimum radiation.

The patient is placed so that the malignant cells are at the centre of the circle in which the cobalt source moves. Do you see why it is necessary to rotate the source? And why it is necessary to position the patient very carefully?

It is also possible to give treatment by injecting the radioactive source into the patient. Iodine-131 is injected to treat a tumour in the thyroid gland.

STERILISATION

Another important medical use of radioisotopes is for sterilisation. Gamma radiation kills most forms of living organisms, including bacteria. The conventional method for sterilisation is heat treatment: dentists and surgeons boil their instruments in water. But some substances, such as penicillin, cannot be subjected to high temperature. Penicillin is therefore packaged before being sterilised and then the package is radiated in order to sterilise it. Needles for injections and other instruments are often sterilised in this way.

It has been estimated that if bedding in hospitals were more regularly sterilised, the length of stay in hospitals might be reduced by 10%. Sterilisation by heat treatment is cumbersome and results in considerable wear and tear. Sterilisation by gamma radiation might be simple by contrast once a suitable sterilisation unit were installed.

Sterilisation might also be of great value in the food industry. By radiation, the storage life of bacon, ham and

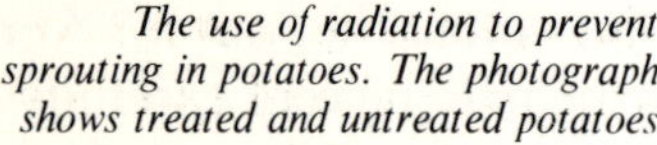

*The use of radiation to prevent
sprouting in potatoes. The photograph
shows treated and untreated potatoes*

sausages, for example, might be increased four to five times. By radiation it is possible to kill or sterilise grain weevils and beetles in grain stores; very great losses have been avoided wherever this has been done. The storage life of potatoes has been considerably increased; potatoes start shooting when stored and this can be prevented by irradiation.

USES OF RADIOISOTOPES IN AGRICULTURE

Radiation from a radioisotope can increase the natural genetic mutation rates in plants and enable breeding to be speeded up. It has also made possible the development of strains of plants with new characteristics. For example, wheat has been produced with shorter straw length, and oats with particular resistance to disease. It is likely that important advances will be made in the next few years, advances of special significance in view of the world food shortage.

Radioisotopes make their contribution as tracers to agricultural work. For example, wheat takes up phosphorus, partly from the soil and partly from the fertiliser which is added. In order to find out the proportions, radioactive phosphorus can be used in the fertiliser so that it can be easily traced.

Cattle food normally has vitamins added to it. There is, however, a very small proportion of vitamin in a large amount of food. Mixing is therefore very important: the supplier of cattle food needs to assure the farmer that there is the right proportion in each sack. By adding a short-lived radioisotope to the vitamin supply, each sack can be tested to see that it contains the right amount after mixing.

An interesting example of the use of radioisotopes was the method used to eliminate the very serious menace of screw-worm fly in the island of Curaçao. Insecticides only reduced the number of these flies to a low level and did not exterminate them. As soon as the use of expensive insecticide stopped, the flies multiplied again very rapidly indeed. To supplement the effect of the insecticide, a large number of male flies were bred and then sterilised (in the breeding sense) by radiation. These were then released amongst the fly population, already weakened by the insecticide. The female screw-worm fly only mates once and so by providing this large number of sterile males the population was further reduced. Provision of more sterile males finally removed the pest from the island.

CARBON DATING

The radioisotope carbon-14 has proved to be of particular interest to archaeologists. In the atmosphere there are neutrons which have been produced by cosmic rays. These neutrons may interact with the nitrogen nuclei present in the air and produce the radioisotope carbon-14 by the reaction:

$$^{1}_{0}n + ^{14}_{7}N \longrightarrow ^{14}_{6}C + ^{1}_{1}H$$

The carbon soon combines with oxygen to form carbon-14 dioxide and this is absorbed by plants together with the common carbon-12 dioxide.

Living wood therefore always contains a certain proportion of carbon-14 and this amount is found to be constant. But as soon as the tree dies, it no longer takes in carbon-14. The carbon-14 already there decays without replenishment. It emits a beta particle and has a half-life of 5600 years.

By examining the proportion of carbon-14 to ordinary carbon in the wood taken from Egyptian tombs, it is possible to estimate their age (assuming of course that the proportion of carbon-14 to ordinary carbon was the same in living wood in Egyptian times as it is today).

As ancient manuscripts were made of papyrus, derived from wood, they may be dated by this method – or at least we may estimate when the material was last in living plant form. This technique has become of considerable importance to archaeology.

USES OF RADIOISOTOPES IN INDUSTRY

To measure the amount of wear in moving parts of an engine, a radioactive piston ring, for example, is put in it (the activity is produced by neutron bombardment in a reactor). Some of the metal will be rubbed off in use and, as it contains a radioisotope, can be detected in the oil without dismantling the engine. This form of testing takes hours instead of weeks by more conventional methods. It makes improved lubrication possible.

Radioisotopes can be used as tracers in studying the movement of sand and mud in rivers and ports. The radioisotope ^{46}Si 'labels' the sand and its movement can be detected by lowering Geiger counters to the seabed. This information can lead to more effective and more economic dredging operations.

There can be a considerable danger of fire when inflammable liquids flow through pipes and build up an electric charge. This industrial hazard can be avoided by positioning radioactive sources which prevent the charge building up. The source ionises the air and lets the charge leak away, just as the radium source in your school laboratory caused an electroscope to discharge.

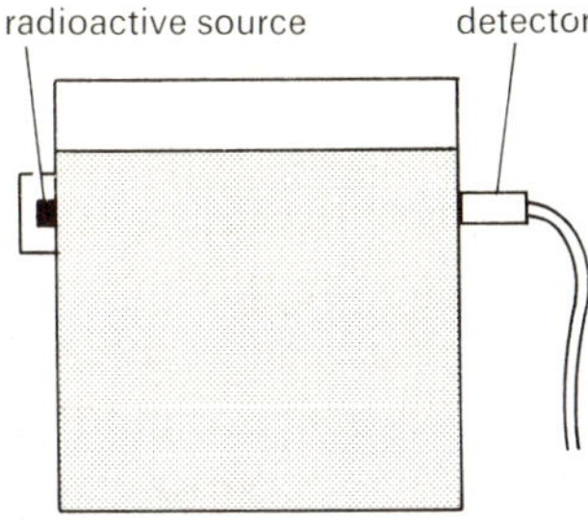

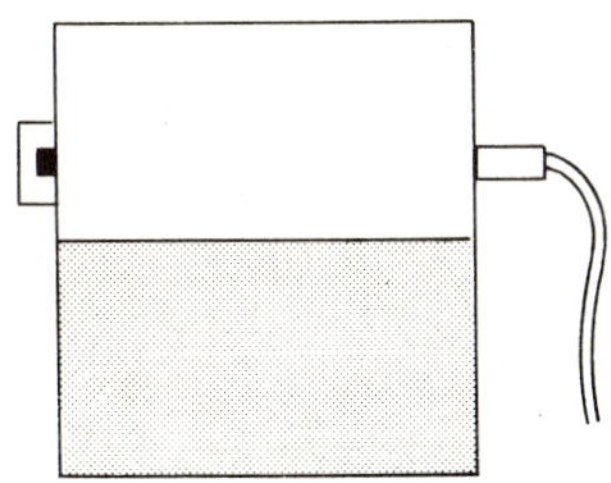

The elimination of charge is also important when manufacturing celluloid and similar materials and when weaving with synthetic fibres. Dust is attracted by charge and dirties the material. This became a very serious problem in the manufacturing of fabrics, especially when the looms were turned off at night: there was always a dirt mark the next morning, which was extremely costly and difficult to remove. A beta source, however, removes the charge and thereby saves the industry unnecessary loss. Unfortunately there is a health hazard with beta sources, and alpha sources (for example, foils of ^{241}Am) have been used instead.

By measuring the amount of radiation which passes from one side to the other we can find the level of a liquid in a closed container. In the first diagram on the left, the liquid comes between the source and the detector and the count-rate is low. But as soon as the level falls, as shown in the second diagram, the count-rate increases considerably.

The same principle enables a manufacturer to check on the amount of toothpaste in a tube, soap powder in a packet, tablets in a box and many similar examples. It is possible to arrange for the automatic rejection of the packet as soon as the detector shows that there is too little

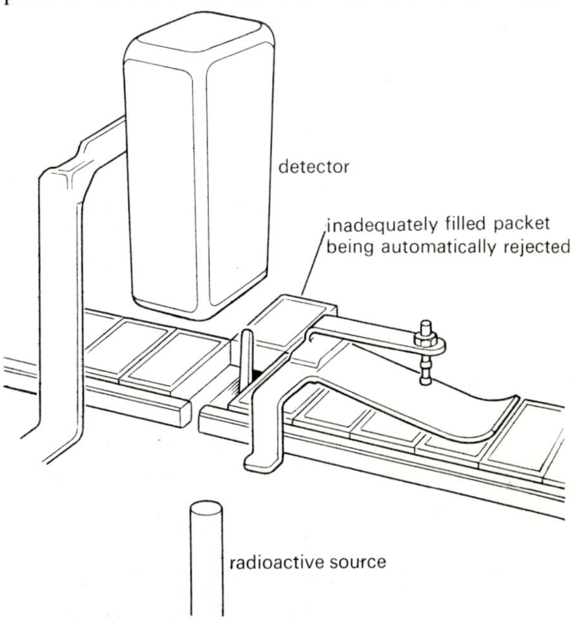

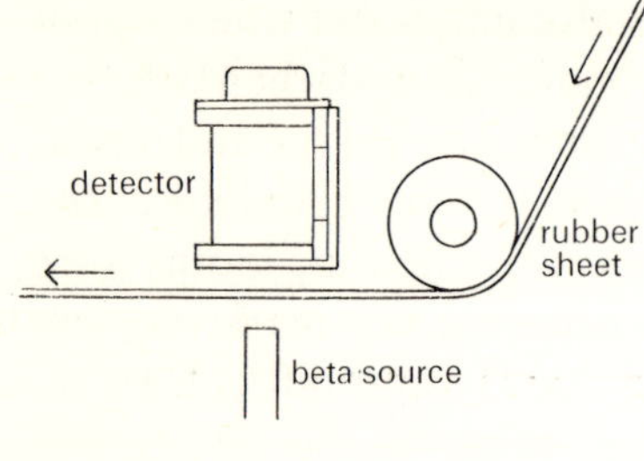

material in the package. This automatic checking permits greater speed and efficiency in production, and hence lowers the cost.

Automatic thickness measurement is made possible by the use of radioisotopes, without having to stop production and touch the material to make measurements by traditional means. The material passes between a radioisotope and a detector: the count-rate will increase if the material is too thin; it will decrease if it is too thick. Not only can this measure the thickness, but it also permits automatic control. If the thickness is too great, the pressure exerted by the rollers can automatically be increased to correct the thickness. Beta radiation can be used in this way for thicknesses ranging from tissue-paper to 2-mm steel, and gamma radiation for steel up to 10 cm.

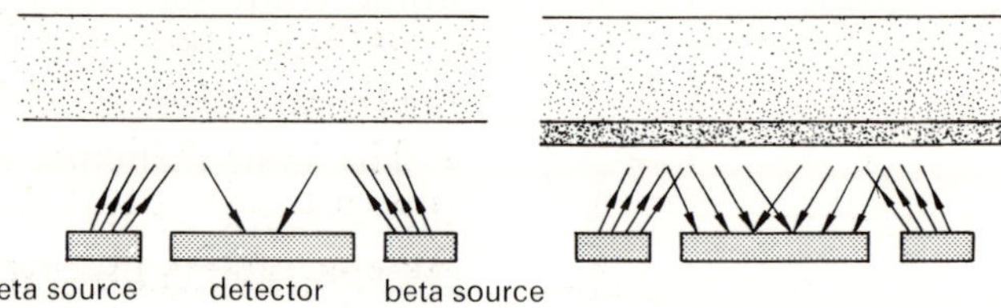

As beta particles are more easily scattered back from surfaces made of materials of high atomic number, it is possible to use radioisotopes for testing the thickness of paint or other coatings. In the arrangement shown above, much more beta radiation is scattered back when the coating of paint, which has a lead content, is thick, less when the thickness is thin or non-existent. Again this leads to automatic control and thus to lower costs of production.

Gamma rays can also be used in place of X-rays to detect flaws in metal welds and castings. The photograph left shows a radiograph taken with a cobalt-60 source.

If X-rays were to be used with thick steel, it would require a several million volt supply to give them sufficient penetrating energy. A cobalt-60 source is very much cheaper and has an equivalent radiation. The photograph on the bottom of the opposite page shows a cobalt-60 source being put in position to photograph a welded seam in a 76-mm thick steel boiler drum.

This is typical of the economies that are made possible

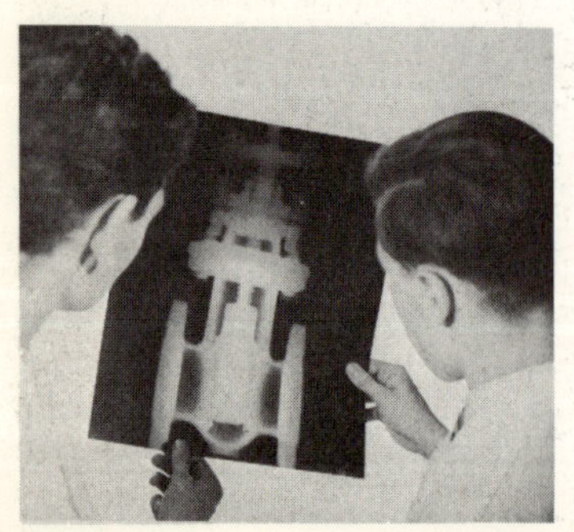

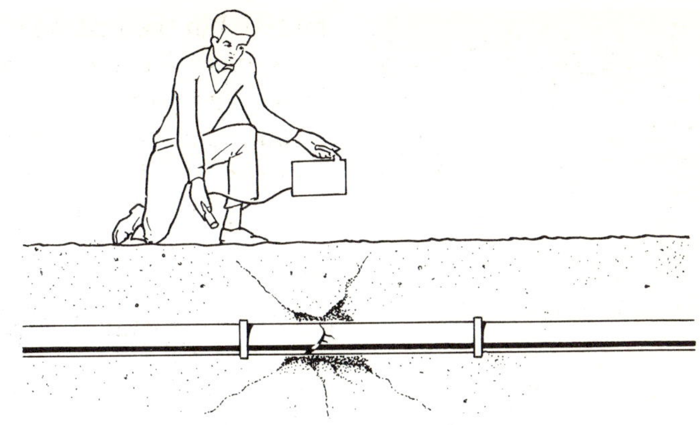

by using radioisotopes and there are many more examples. Suppose a leak develops somewhere along a pipeline buried beneath the ground. By traditional methods it would have been necessary to dig up the pipe, perhaps over many miles, to find the leak. Now we can send a solution of a suitable gamma emitter (^{24}Na) through the pipe, and detectors moving along the ground over the pipeline will show where some of it has leaked.

Cobalt-60 source being put into position to photograph a welded seam in a boiler drum

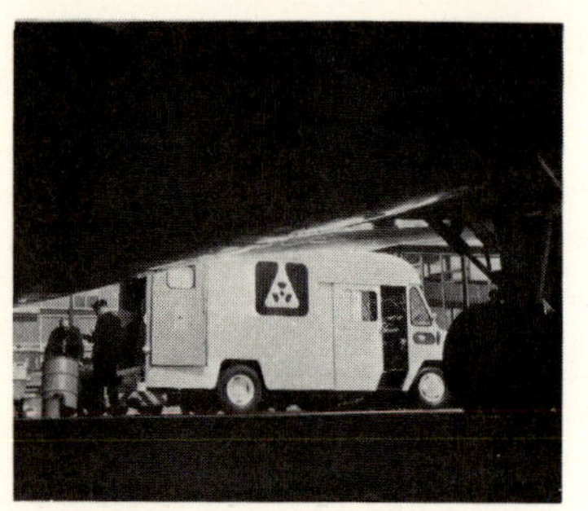

A consignment of radioisotopes being loaded into a plane

SUPPLY OF RADIOISOTOPES

The choice of radioisotope for a particular application will depend first on what penetration is necessary, then on how strong the source needs to be and finally on how long the activity should last. If the half-life is too short, the isotope may have decayed too much before the experiment has begun; on the other hand, in some experiments, for example when finding the degree of mixing in cattle food, it is important that the half-life should be short enough for the activity to have ceased by the time the cattle eat the food, and of course the decayed material must not be poisonous. There are already over a thousand different radioactive isotopes and the right one has to be chosen for the particular application.

Radioisotopes are now sent all over the world from the Radiochemical Centre at Amersham in Buckinghamshire. Safety is important and the radioisotopes are sent in special containers. Speed is also essential, especially for short-lived isotopes. The despatch of radioisotopes has become so much a matter of routine that almost all British planes have special arrangements for transporting them.

In 1969, the Radiochemical Centre sent out over 80 000 consignments of radioisotopes; 57% of these went overseas.

CONCLUSION

From the earliest experiments which Becquerel performed in 1897 to the present day is a mere seventy or so years. In that time, man's understanding of the structure of matter and of the atom itself has developed to such a pitch that we can derive energy from controlled nuclear reactions; we can heal; we can use atoms themselves to help us to understand complex natural processes. But we have also gained the power to destroy whole cities and devastate entire countries, perhaps even life itself. If men are to remain masters of their own achievements, and to use them for good rather than evil, understanding and wisdom will both be essential.

These notes are a discussion of the various numbered questions in the text.

1. The photograph shows alpha particles from a polonium source passing through a cloud chamber. The interesting feature is that the alpha particles in air under normal conditions all have roughly the same range; this is a characteristic of alpha particles, as you probably saw with a spark counter.

On the other hand, the lengths of the tracks are not *exactly* the same; there is some *straggling*, as it is technically called. The range, however, does depend on the energy of the alpha particle; it comes to rest when it has lost all its energy and the loss of energy will depend on the number of collisions on its way. This will show statistical variations: even though all the alpha particles set out with the same energy, we would not expect all to have the same range.

2. The photograph shows alpha particles with two distinct ranges: this is because the source contains two different radioactive substances.

3. There are fewer tracks visible in the second photograph than in the first as the radioactive gas has 'decayed'. In fact the half-life of thoron is 54 seconds, so that after two minutes we would expect the *activity* to be about a quarter.

You may also have noticed that the length of the tracks appears greater after there has been decay in the chamber. This is not due to the alpha particles, but merely to a characteristic of the cloud chamber. When an alpha-particle track has been formed by condensation of water droplets, the water vapour in the vicinity is momentarily used up. If another alpha particle crosses the same path, there is not enough water vapour so the track appears to stop, even though the alpha particle obviously must continue on its way. So, when there are a lot of alpha particles moving in all directions (as in the first of the two photographs under discussion) the tracks will appear shorter than they do when there are fewer (as in the second photograph).

4. The first track is characteristic of the alpha particle and is like all the other tracks you have seen so far. The

second, however, is more 'wispy'; there is not the same density of water droplets. The third has a very similar appearance to the second, but it is a tortuous, curved path. Both these are in fact caused by beta particles: the first has high energy, whereas the second is a low-energy particle, obviously much buffeted in its path. The fourth shows the path of a gamma ray. It has not made a visible track itself; you notice its path by the tracks of the secondary electrons it has produced on its journey by ionisation of the air molecules. These secondary electrons produce visible tracks even though the gamma ray did not.

5. Alpha and beta particles would be deflected in opposite directions in a magnetic field because they are oppositely charged: alpha particles are positive, beta particles negative. As gamma radiation does not consist of charged particles (it is electromagnetic radiation), it is undeflected in a magnetic field.

Madame Curie was the first to use the diagram on the left, representing a radioactive source in a lead container, to show alpha, beta and gamma radiation passing through a magnetic field at right angles to the page. The alpha particles are deflected a small distance, the beta particles much more in the opposite direction and the gamma radiation is undeflected. In fact this experiment was never done and could not be done like this: alpha particles are much too difficult to deflect.

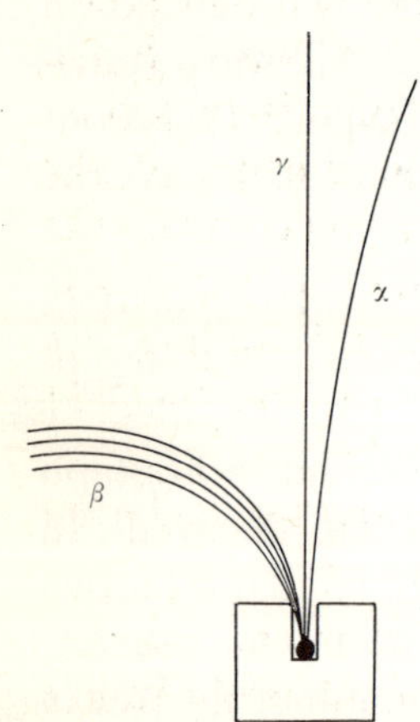

6. The slower-moving particle is easier to deflect than a fast one.

7. The electron is gradually losing energy because of the ionisation it is producing. As its velocity falls, so the curvature increases and the circles become smaller.

8. Beta particles or electrons. Alpha particles would have denser tracks.

9. The electrons crossing the photograph must have much higher energies as they are deflected much less.

10. As soon as the distance of the source from the counter exceeded the range of the alpha particles in air, the sparks ceased. This is a simple way of measuring the range.

11. The sparks stopped as soon as a piece of paper was

inserted between the source and the spark counter. Alpha particles cannot penetrate through paper.

12. The sparks were random in time: sometimes they were close together, sometimes there was a longer interval between them.

13. Many luminous watches give off alpha and beta particles, of which the beta particles only pass through the glass. If the face of such a watch is held against the thin end window of a GM tube, quite high count-rates may be noticed. The count-rate is very much less if the end window is held against the back of the watch or if a thick-walled GM tube is used.

14. Even without any radioactive sources near, the scaler and GM tubes will record a 'background count'. Some of this comes from our surroundings, some is caused by radiation from outer space. A background count of fifty per minute would not be unusual. When you compare the area of surface of the tube with the area of your body, you can estimate how much radiation is falling per minute on you. That calculation assumes that the GM tube is 100% efficient and that it counts every bit of radiation falling on it: in fact it is not, so the radiation hitting you every minute is even higher.

15. As beta particles are electrically charged, they are deflected in the opposite direction if the magnetic field is reversed. In the experiment under discussion, they would be deflected away from the GM tube and not toward it.

16. A television tube.

17. Increasing the filament current would increase the number of electrons given off. It would therefore increase the quantity of X-radiation given off. Increasing the accelerating voltage would give the electrons greater energy when they hit the anticathode: X-rays with higher frequency (greater energy) would be given off. Increasing the accelerating voltage would therefore produce more highly penetrating radiation.

ACKNOWLEDGEMENTS The authors and publisher are grateful to Taylor & Francis Ltd for permission to reproduce the extract on pages 31–2 from the *Philosophical Magazine*, Volume XVII.

We are also grateful to the following for permission to reproduce photographs: page 4 (*above*) Keystone Press; front cover and pages 4 (*below*), 52, 54, 59 and 60 United Kingdom Atomic Energy Authority; page 6 (*above*) Ministry of Defence (Crown copyright reserved); pages 6 (*below*), 7 (*right*), 9, 10 (*left*), 11 (*above*), 22 (*top*), 24 (*above*), 27, 38 (*above right*), 47 and 48 Science Museum, London (Crown copyright reserved); pages 7 (*left*), 10 (*above and below right*), 11 (*below*), 37 and 50 Pergamon Press Ltd, from *An Atlas of Typical Expansion Chamber Photographs* by Gentner, Maier-Leibnitz and Bothe; page 8 Bernard Taylor, from 'The Taylor Cloud Chamber', *Technical Education and Industrial Training*, July 1966; pages 13, 15, 21 (*right*) and 42 (*right*) Esso Petroleum Co Ltd; page 14 (*above right*) Philip Harris Ltd; page 14 (*below left*) Panax Equipment Ltd; pages 20 and 30 Cavendish Laboratory, University of Cambridge; page 22 (*centre and below*) The Faculty of Radiologists Film Library; page 23 Mullard; page 24 (*below*) Radio Times Hulton Picture Library; page 25 the Master and Fellows of Sidney Sussex College, Cambridge; page 28 Griffin & George Ltd; page 29 CERN; pages 38 (*above left*) and 46 (*below left*) Lord Blackett; pages 38 (*below right*), 40 and 42 (*left*) E. J. Wenham; page 51 J. L. Lewis; page 58 Central Office of Information (Crown copyright reserved).

We should also like to acknowledge the following sources of photographs: page 7 (*left*) J. K. Bøggild; page 10 (*above right*) P. I. Dee, *Proc. Roy. Soc.*, London (A) **136**, 727 (1932); page 11 (*below*) Radiation Laboratory, Berkeley, California; page 37 F. Joliot, *J. Phys. Radium* **5**, 219 (1943); page 50 I. K. Bøggild, *K. Danske Vidensk Selsk. Mat.-fys. Medd.* 18 (1940).